California Natural History Guides: 41

Edible and Useful Plants of California

by Charlotte Bringle Clarke

UNIVERSITY OF CALIFORNIA PRESS
BERKELEY • LOS ANGELES • LONDON

California Natural History Guides
Arthur C. Smith, General Editor

Cover Photographs:
Upper left: Elderberry—*Sambucus mexicana* (p. 40)
Upper right: Sow Thistle—*Sonchus oleraceus* (p. 224)
Lower left: Morel—*Morchella esculenta* (p. 62)
        (Photo courtesy of U.C.L.A.)
Lower right: Wood Strawberry—*Fragaria vesca* (p. 91)
        (Photo courtesy of R. M. Hubert)

PHOTOGRAPHIC CREDITS: Plate 1c by L. DeBuhr; Plates 2a, 2d, 4a, 4e, and 8f
by Mary Hood; Plate 3c by R. M. Hubert; Plate 5b by T. Bandaruk; Plate 5e
by Dr. Wheeler North. All other photos by the author.

DRAWING CREDITS: *Macrocystis pyrifera* by H. R. Williams; *Taraxacum
officinale,* map of California counties (adapted from Munz, 1959), and
Glossary illustrations by James E. Clarke. All other drawings by Karlin
Grunau.

University of California Press
Berkeley and Los Angeles, California
University of California Press, Ltd.
London, England

Copyright © 1977, by the Regents of the University of California
ISBN: 0-520-03261-6
Library of Congress Catalog Card Number: 76-14317
Printed in the United States of America

To Oscar F. Clarke,
who sowed the seeds from which I reap.

ACKNOWLEDGMENTS

I am indebted to the many students and colleagues who urged me to write this book, but especially to my dear friend Oscar Clarke of the University of California, Riverside, who participated in many phases of its preparation. I am also indebted to Dr. Mildred Mathias of the University of California, Los Angeles, who reviewed the manuscript's rough draft and made many valuable suggestions, as well as to Dr. Robert F. Thorne of Rancho Santa Ana Botanic Garden for assistance with taxa. For her dedication to detail and fresh specimens, I would like to thank my artist, Karlin Grunau, who many times took the initiative to get things accomplished without my supervision. For reviewing the fungi included herein, I wish to thank Dr. Martin Stoner of California Polytechnic University, Pomona, and Mr. Wayne Bebaut. For assistance with seaweeds, Tony Stagnalini, and for manuscript preparation, Mandy Cole; I am grateful for the help of both. I am also grateful for the information set forth by many of the authors listed in the last chapter of this book and to the following people who have helped me either morally or monetarily in my career: Milo and Zella Barber, Professor Willard Barber, Dr. Gerald Sjule, and my patient and helpful husband, James.

CONTENTS

INTRODUCTION

In great part, this book has arisen out of popular demand. The "back-to-nature" movement, which has taken such firm root in our country in the last ten years, has kindled the desire to return to the ways of our forefathers. The interest in all things natural, including good nutrition and man's relationship with his environment, has increased along with rising food costs and pollution, and many people are searching to rediscover the knowledge of the Indians and early settlers concerning wild plants.

The uses of California plants have heretofore been scattered in journals and various publications not usually available to the general public. This book expands on the existing sources and tells about both aboriginal and modern uses of more than 220 different species of plants found in California. Recipes are given where applicable. I have written this book for the person with no knowledge of botany, and it is meant to be useful to the average student, city dweller, backpacker, camper, vegetarian, and survival instructor—be they trained in botany or not. For this reason, the plants are arranged by habitat, and within each habitat they are listed alphabetically by the most frequently used common name. When looking up a plant, you should refer to the General Index first, because many plants have several common names and might be listed under a different name than the one you are familiar with. I hope you will also refer to the Index to Plant Uses which lists plants by their parts and products. Extensive cross-references and a glossary are also included for your convenience.

Although the book contains many recipes that call for refined sugar and flour, there is none in which brown sugar and unbleached flour cannot be substituted if you prefer the more "natural" approach. However, in many recipes, substituting honey for sugar changes the flavor or consistency so that it is not advisable to try this. (If you do experiment, substitute seven-eighths cup of honey or molasses for each cup of sugar, then cut the liquid in the recipe by one-quarter cup for each cup of honey used. You should also reduce the baking temperature by 25° F to prevent overbrowning. Please inform me of the results by writing in care of the publisher.)

Nutritional information has been given when it is known. Small amounts of wild foods are usually called for in the recipes (for instance, cups instead of quarts) because I realize that many foragers do not have the time to accumulate large quantities of wild plants. Most of the recipes can be doubled or tripled.

I have been teaching edible plant courses for a number of years, and often receive the same questions from different students. The most frequently asked questions are discussed below.

Is a particular plant edible or poisonous? There are few guidelines I can offer to help the reader determine edible from poisonous plants, but *you should never eat a plant that you have not positively identified.* Following this rule, you should start with a recipe calling for a plant you know for certain and then branch out plant by plant as you acquire techniques of identification. Positive identification is not difficult and I discuss it in the last chapter, Obtaining Further Assistance.

The edible plants described in this book are not just edible, but palatable. Many plants listed in ancient journals as edible were used only in emergencies and would not generally appeal to popular tastes today. To make this book a convenient size, I have included only the most delicious, abundant, and easily identified plants. For more extensive lists, refer to the Suggested References.

How nutritious are wild plants? Perhaps because of their slower growth rate, wild plants have been shown to be consistently higher in vitamins and minerals than cultivated plants. They have not been developed for their color or storage qualities—they are natural and fresh. Most are unsprayed and unprocessed and have a natural waxy coating which prevents dirt from getting inside their tissues. Therefore, a simple washing is often all that is needed to prepare them for cooking. One exception to this can occur in vacant lots that have been sprayed with weed killers; if the foliage of plants looks brown, withered, and oily, they should not be used for food. My best foraging in the city is in vacant lots that have been plowed, but not sprayed. Because of the lead content of some gasolines, it is wise to gather plants at least 100 feet from a busy road.

Why is a plant bitter when a reference says it is edible? Like most vegetables, wild plants must be picked at the right time and treated properly. If you have ever tasted asparagus that has grown too tall, or garden lettuce that has gone to seed, you have experienced how tough and bitter they can become. Wild plants are similar. Many are unpalatable when eaten raw, just as raw eggplant or rhubarb would be. Cooking is sometimes necessary to rid plants of water-soluble properties that are bitter or even poisonous (in the case of rhubarb). The common Dandelion is a good example. If it has matured enough to flower (if it is in your lawn and frequently mowed, it most certainly has), it is quite bitter. Before maturation, however, the leaves are tasty additions to many dishes, raw or cooked. When they are bitter, they are really bitter, and require two boilings to make them mild again. Boiling, incidentally, destroys only about one-third of the vitamins, leaving plenty behind.

Can I save money using wild plants for food? Gathering wild plants is best enjoyed when approached with the idea of having fun rather than saving money. And it does require time! Although many plants need only to be gathered, washed, and thrown in the pot, others must be hulled, soaked, or prepared in some time-consuming way. But nutting, berrying, and general foraging can become wonderful family adventures, and if enjoyed in this light, the food products which result are certainly saving you money. Only you can decide what your time is worth and how you wish to spend it. Personally, I would rather forage than tend a garden, which requires mulching, irrigating, weeding, and fertilizing.

I want to try this plant, but it looks so strange! Prejudice, especially in those who have been raised in the city, is a fact. I sidestep this by preparing plants in a tasty, though somewhat disguised, dish whenever possible. My guests can then let their taste buds be their guide instead of preconceived notions of what is and what is not good food. Perhaps this would work for you. Give the plant an honest trial. If you do not like it raw, try it boiled or baked. If you still do not care for it, check to see if you are using the right part at the right time. With potherbs, for example, usually only the young leafy greens are palatable.

Where do I start? To begin learning about wild plants, it is helpful to know what is available at a particular time of the year. Generally, underground parts of plants are available all year; nuts and berries develop in the summer and fall; while other fruits and herbs are found in the spring. Weeds (bless them!) can be found the year around. If you cannot decide where to start, feel free to write me in care of the publisher and I will help if I am able. For proper plant identification, see the last chapter, Obtaining Further Assistance.

A word of caution—great care has been taken in this book to treat only those plants which are easily found, abundant, and identifiable. Since some people have food allergies of which they are unaware, it is wise to eat wild food in moderation the first time, however delicious. The ecology-conscious reader will also heed my frequent appeals to limit foraging to those plants which are particularly abundant and not destroyed by gathering. Occasionally a near relative of a plant treated here is in an endangered state. In such cases I caution you to refer to the Inventory of Rare and Endangered Vascular Plants of California published by the California Native Plant Society. Please heed these cautions. In addition, there are laws that technically protect all plants on public lands from destruction, but gathering of fruits and nuts under the supervision of the United States Forestry Service is generally permissible. You should check with the local authorities before foraging and take care to leave plenty of plants for the animals that depend upon them for food.

The taxonomic authority used in this book is P. A. Munz's *A California Flora,* as supplemented in 1968, for native plants. L. H. Bailey's *Manual of Cultivated Plants* has been used for ornamentals. Recent taxonomic changes also have been added by Drs. Mildred Mathias and Robert F. Thorne. All voucher specimens relating to each species discussed here were confirmed at the Rancho Santa Ana Botanic Garden in Claremont, California, or the Herbarium of the University of California at Los Angeles.

The information in the distribution and habitat sections of this book is patterned after previous California Natural History Guides. I would like to give a brief summary of the plant communities. Beginning at the high tide line of the Pacific Ocean, the sand dunes are the basic land form that supports the *Coastal Strand* vegetation. Sporadically along the northern California coast the *Coastal Prairie* community is found, just inland from the Coastal Strand, distinguished by extensive grasslands. In estuaries, bays and other areas that are protected from wave action and strong winds the *Coastal Salt Marsh* is found, with typical plants in this association being Glasswort or Pickleweed species of the genus *Salicornia*. Continuing in an eastward transect, in northern California immediately inland from the Coastal Prairie is a plant community characterized by the presence of shrubs intermixed with grassy meadows, called the *Northern Coastal Scrub* community, with Salal (*Gaultheria shallon*) as a typical member. Further inland from the immediate coast is the *Closed-Cone Pine Forest,* where coniferous (cone-bearing) trees grow, such as the Bishop Pine (*Pinus muricata*), Monterey Pine (*Pinus radiata*), and various cypresses. The *North Coastal Forest* includes the plants that occur in the Redwood Forest, Douglas Fir Forest, Mixed Evergreen Forest, and North Coastal Coniferous Forest, typical trees of these regions being firs, maples, oaks, hemlocks, and redwoods. *Chaparral* refers to the almost impenetrable foothill vegetation that includes an abundance of large shrubs, such as Chamise (*Adenostoma fasciculatum*), Manzanita (*Arctostaphylos* ssp.), California lilac (*Ceanothus* ssp.) Scrub Oak (*Quercus dumosa*), and yucca.

Large areas of the valleys and eastern slopes of the North and South Coast Ranges, the valleys of interior southern California, and the western foothills of the Sierra Nevada are occupied by the *Valley and Foothill Woodland* plant community. This plant community ranges from 300-5000 ft (100-1500 m), reaching its highest elevations in southern California. This community is characteristically dominated by Oak trees with an understory of shrubs or grasses. The Valley and Foothill Woodland includes plants which are often listed under Northern Oak Woodland, Southern Oak Woodland, and Foothill Woodland. The *Valley Grassland* plant community occupies regions of the Central Valley and some areas of the Coast, Transverse, and Peninsular Ranges. Today it is characterized by introduced grasses, and in the spring by its wildflower displays.

The *Riparian Woodland* plant community is represented by trees, shrubs, and herbs that are restricted to the banks of water courses; Bigleaf Maple (*Acer macrophyllum*), Black Cottonwood (*Populus trichocarpa*), California Sycamore (*Platanus racemosa*), Fremont Cottonwood (*Populus fremontii*), and willows (*Salix* ssp.) are typical plants of this community. The *Freshwater Marsh* occurs in areas where there are expanses of standing or slow-moving shallow water; cattails (*Typha* ssp.), bulrush or tule (*Scirpus* ssp.), and sedge (*Carex* ssp.) are typical plants of this community. A *Montane Forest* is a higher (mountain) plant community which is also known as *Yellow Pine Forest*. It is characterized by White Fir (*Abies concolor*), Incense Cedar (*Calocedrus decurrens*), Giant Sequoia (*Sequoiadendron giganteum*), and numerous pines, including the Ponderosa Pine (*Pinus ponderosa*). Above the Montane Forest is the *Subalpine Forest*. At its lowest, it occurs at 5000 ft (1500 m) and extends to 11,000 ft (3400 m) in some areas of the desert (e.g., White Mountains) and Southern California. The Subalpine Forest is dominated by conifers, especially the Lodgepole Pine (*Pinus murrayana*). The harsh higher elevations above the timberline generally support only dwarfed plants that often exhibit a cushion-like growth. This is the *Alpine Fell-field* plant community. On the eastern slopes of the Sierra Nevada, below the crest of the mountains occurs the *Pinyon-Juniper Woodland*. This community takes its name from the Single-leaf Pinyon Pine (*Pinus monophylla*) and various junipers. Often adjacent to the *Pinyon-Juniper Woodland* is the

Sagebrush Scrub community named after the Basin Sage-brush (*Artemisia tridentata*).

There are several additional plant communities that are found only in southern California. *Coastal Sage Scrub* is the southern counterpart of the *Northern Coastal Scrub*. It occupies a narrow strip along the coast and gets its name from the presence of several species of *Salvia,* such as *Salvia mellifera,* Black Sage. Other common shrubs include Coastal Sagebrush (*Artemisia californica*), Wild Buckwheat (*Eriogonum fasiculatum*), and Lemonade Berry (*Rhus integrifolia*). A desert plant community dominated by small, grayish-leaved shrubs is the *Shadscale Scrub.* Mormon tea (*Ephedra* ssp.) is found here as well as Shadscale (*Atriplex confertifolia*). Low-lying, poorly drained alkali flats found in the low deserts are made up of shrubs that belong to the Goosefoot Family (Chenopodiaceae) whose members are salt-tolerant They form the *Alkali Sink Scrub.* Well-drained mesas and desert slopes support the *Joshua Tree Woodland,* dominated by the familiar Joshua Tree (*Yucca brevifolia*), junipers, mormon tea (*Ephedra* ssp.), and cholla cactus (*Opuntia* ssp.). The most widespread desert plant community in southern California is the *Creosote Bush Scrub.* Creosote Bush is *Larrea divaricata.*

For more details on these associations, see *Introduction to California Plant Life* by Ornduff, listed in the references.

The information regarding animal uses of foods has come from many sources, but I have relied heavily upon *American Wildlife and Plants* by Martin, Zim, and Nelson, listed in the references.

The recipes are my own and have been invented outright or adapted from existing recipes for domestic or wild plants. They have all been tested and judged by friends, dinner guests, or students. I would appreciate any comments or corrections from readers who try them. I am especially interested in new variations which might increase the enjoyment of any of the plants described here, and it is my sincere hope that this book will increase your knowledge and appreciation of our wonderful flora and the people who have found it useful.

COUNTIES OF CALIFORNIA

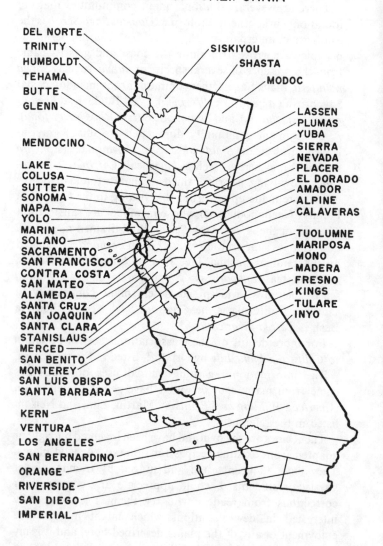

DEL NORTE
TRINITY
HUMBOLDT
TEHAMA
BUTTE
GLENN

SISKIYOU
SHASTA
MODOC

LASSEN
PLUMAS
YUBA
SIERRA
NEVADA
PLACER
EL DORADO
AMADOR
ALPINE
CALAVERAS

MENDOCINO

LAKE
COLUSA
SUTTER
SONOMA
NAPA
YOLO
MARIN
SOLANO
SACRAMENTO
SAN FRANCISCO
CONTRA COSTA
SAN MATEO
ALAMEDA
SANTA CRUZ
SAN JOAQUIN
SANTA CLARA
STANISLAUS
MERCED
SAN BENITO
MONTEREY
SAN LUIS OBISPO
SANTA BARBARA

TUOLUMNE
MARIPOSA
MONO
MADERA
FRESNO
KINGS
TULARE
INYO

KERN
VENTURA
LOS ANGELES
SAN BERNARDINO
ORANGE
RIVERSIDE
SAN DIEGO
IMPERIAL

ILLUSTRATED GLOSSARY

Acute—Tapering to the tip (apex), with the sides straight or nearly so.

Alternate—Borne singly and not opposite—in leaves one at a node. (Fig. ii.)

Annual—Completing the life cycle in one growing season.

Anther—The pollen-bearing part of the stamen. (Fig. i.)

Awn—A slender bristlelike organ usually at the tip of a structure.

Axil—The angle between an organ and its axis.

Axillary bud—A bud found in the angle formed between the stem and leaf. (Fig. ii.)

Basal—Relating to, or situated at, the base.

Biennial—Of two years' duration from seed to maturity and death.

Blade—The expanded portion of a leaf. (Fig. ii.)

Bloom—As used here, most often refers to a whitish coating on a plant.

Bract—A more-or-less modified leaf situated near a flower or inflorescence.

Burr—A seed or fruit bearing spines or prickles, these often hooked or barbed.

Calyx—The outer series of flower parts. Collective for sepals. (Fig. i.)

Cambium—A meristematic zone (area of cell division).

Capillary plume—Slender, feathery structure, such as the modified calyx in the Compositae.

Catkin—A spikelike, usually pendulous inflorescence of unisexual flowers.

Clasping base (in reference to leaves)—A sessile leaf with the lower edges of the blade partly surrounding the stem.

Clawed (petal)—Having a narrow base or stalk.

Compound (leaves)—A leaf completely separated into two or more leaflets. (Fig. ii.)

Corolla—The inner series of flower parts. Collective for petals. (Fig. i.)

Crenate—Toothed, with teeth rounded at the apex. (Fig. iii.)

9

FLOWER PARTS

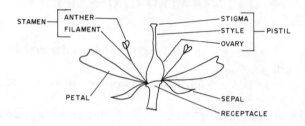

INFLORESCENCES

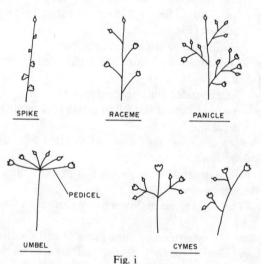

Fig. i

Cyme—A flower cluster, often convex or flattopped, in which the central or terminal flower blooms earliest. (Fig. i.)

Deciduous—Falling away, not persistent or evergreen.

Dentate—Toothed, with the teeth directed outward. Sometimes loosely used for any large teeth. (Fig. iii.)

Drupe—A fleshy, indehiscent, one-seeded fruit, the inner layer stony.

Elliptic—Shaped like an ellipse—widest in the center with the two ends equal. (Fig. iii.)

LEAF ARRANGEMENT

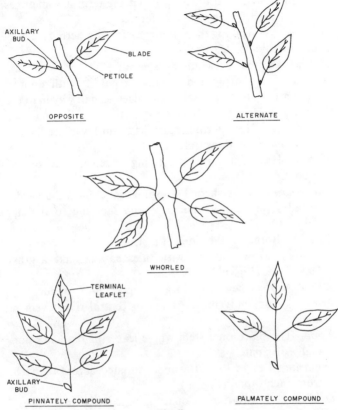

Fig. ii

Entire—Margins without teeth or lobes, such as a smooth leaf margin or edge.

Exfoliating—Peeling off in thin layers.

Exserted—Projecting beyond a surrounding organ, as a stamen exserted from a corolla.

Filament—The thread-like part of the stamen that supports the anther. (Fig. i.)

Foliate—Having leaves.

Gill—The knife-blade-like structures on the underside of the cap of a mushroom.

12 *Illustrated Glossary*

Glandular—Bearing glands. A glandular hair has an enlargement like a hatpin at its tip.

Glaucous—Covered with a whitish or bluish waxy covering, which should readily rub off; but the term is sometimes loosely used for any whitish surface.

Herbaceous—Having the characteristic of an herb; also leaf-like in color or texture.

Hybrid—A cross between two species.

Incised—Cut sharply and usually irregularly with sinuses, deeper than teeth but seldom as deep as half way in to the base or midrib of the leaf. (Fig. iii.)

Indehiscent—Not splitting open, as a sunflower seed.

Inflorescence—A flower cluster. (Fig. i.)

Keeled—Having a dorsal projecting ridge, like the keel of a boat.

Lanceolate—Lance-shaped—several times longer than wide, broadest at or near the base and tapering to the tip. (Fig. iii.)

Lateral—Borne on the sides of a structure.

Linear—Narrow and flat with sides parallel, like a grass leaf blade. (Fig. iii.)

Lobe—Any segment of an organ, especially if rounded.

Midrib—Also midvein, the main or central rib or vein of a leaf.

Node—The place on a stem where leaves or branches normally originate.

Opposite—Leaves two at a node, situated across the stem from each other. (Fig. ii.)

Orbicular—A two-dimensional figure, circular in outline.

Ovary—That part of the flower that contains the developing seeds. (Fig. i.)

Ovate—Egg-shaped in outline, attached at the wide end. (Fig. iii.)

Palmate—The lobes or divisions attached at or running down toward one place at the base. Also used to describe the veins of certain leaves. (Figs. ii and iii.)

Panicle—A compound inflorescence with the younger flowers at the apex or center. (Fig. i.)

LEAF SHAPES

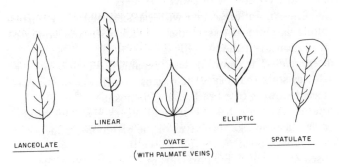

LINEAR

LANCEOLATE

OVATE
(WITH PALMATE VEINS)

ELLIPTIC

SPATULATE

LEAF MARGINS

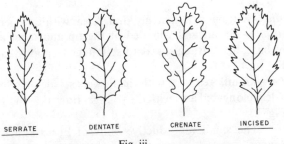

SERRATE   DENTATE   CRENATE   INCISED

Fig. iii

Pedicel—The stalk to a single flower of an inflorescence. (Fig. i.)

Perennial—Lasting from year to year.

Petal—One of the individual parts of the corolla. (Fig. i.)

Petiole—The stalk to a leaf blade or to a compound leaf. (Fig. ii.)

Pinnate—A compound leaf with the leaflets on two opposite sides of an elongated axis. (Fig. ii.)

Pistil—The collective name for the stigma, style, and ovary. (Fig. i.)

Potherb—A leafy green plant normally boiled or steamed for eating.

Prostrate—Lying flat on the ground.

Pubescence—Type of hair. A plant described as pubescent is normally covered with short soft hairs.

Raceme—An inflorescence with pedicelled flowers borne along a more-or-less elongated axis with the younger flowers nearest the tip. (Fig. i.)

Receptacle—The more-or-less expanded portion of the flower stalk that bears the organs of a flower or the collected flowers of a head as in Compositae. (Fig. i.)

Rosette—A dense basal cluster of leaves arranged in circular fashion like the leaves of the common Dandelion.

Scorpioid—Coiled at the apex like the tail of a scorpion, used especially for inflorescences.

Sepal—One of the parts of the outer whorl of the floral parts, or calyx; usually green in color. (Fig. i.)

Serrate—Toothed, with sharp teeth directed forward. (Fig. iii.)

Sessile—Without a stalk of any kind. A leaf without a petiole.

Spatulate—Broad and rounded at the tip and tapering at the base, like a druggist's spatula; flattened- spoon-shaped. (Fig. iii.)

Spike—An inflorescence with the flowers sessile on a more-or-less elongated axis with the younger flowers at the apex. (Fig. i.)

Spore—The small reproductive body of Ferns and lower plants (like mushrooms). Analogous to the seed.

Spur—A hollow, saclike or tubular extension of a floral organ.

Stamen—One of the pollen-bearing organs of a flower, made up of filament and anther. (Fig. i.)

Stellate (hair or pubescence)—Star-shaped.

Stigma—That part of the pistil that receives the pollen. (Fig. i.)

Stipe (in seaweeds)—The stemlike structure that holds up the blades.

Style—The contracted portion of the pistil between the ovary and the stigma in a flower. (Fig. i.)

Succulent—Fleshy and full of juice.

Terete—Circular in cross-section and more-or-less elongated. Similar to cylindrical.

Terminal—Last or end. Referring frequently to the leaflet at the very tip of a compound leaf. (Fig. ii.)

Tuberous—Having a thickened, short, and usually subterranean stem. A potato is a tuber.

Umbel—A convex or flattopped inflorescence, the flowers all arising from one point, the younger in the center. (Fig. i.)

Variable—Having numerous variations. This term used especially with a plant that often does not fit its printed description in many respects.

Whorl—A cluster of three or more leaves or other structures arranged in a circle around a stem or some common axis. (Fig. ii.)

NOTES ON PLANT
PREPARATION TECHNIQUES

Because the average person does not have picking and processing machines at his disposal, I would like to review a few of the basic preparation techniques mentioned in this text.

Leaching. In many instances a water-soluble acid or other distasteful compound can be washed out of a plant by the process of leaching as was done by the Indians in preparing acorns. The basic procedure is to shell the fruits, nuts, or acorns, chop or grind them into bits or meal, and then run water through the meal repeatedly. Boiling water produces the fastest results. When the water turns dark brown, pour it off and add more. The leaching process is finished when the meal is tasted and found free of bitterness. Cold water leaching can be accomplished by placing the meal in a porous cloth sack (such as a pillow case) and letting it lie in a river for a few days. (Some enterprising souls place a sack in the water tank of a toilet and let the flushing operation do the leaching.)

Fruit Extracts. Many wild fruits are delicious but have a thin pulp, or abundant seeds. The easiest way to prepare these fruits is to make an extract of the pulp by simmering the entire fruit, crushed, in water. Let it steep, then strain it through cheesecloth or a jelly bag. This is the best way to extract the flavor from rose hips, for instance. The extract can be boiled down and sweetened for syrup or jelly, or thickened as a sauce.

Drying Fruits and Herbs. Many fruits and herbs may be preserved and even improved in flavor by drying. The easiest way is to lay them out in a very thin layer on cookie sheets or screens and place them in the attic or in a pilot-lit oven for a few days. They must be stirred once or twice a day. If they are totally dried, they may then be kept in clean jars. If only partially dried fruit is desired, it should then be frozen. Thicker fruit, such as mushrooms, should be sliced prior to drying.

Tea Preparation. Many wild herbs make wonderful teas. The amount needed varies with the plant from a few leaves to a handful. Generally speaking, the best teas are made by adding boiling water to the dried plant and then setting it off the fire to steep for about ten minutes. If the fresh herb is used, more plant material will be needed.

Instant Jelly. I have devised a simple way to make small amounts of delicious jelly (or sauce, as you prefer). When the fruit has been prepared (cooked, raw, or soaked), run it through a couple of layers of cheesecloth or a jelly bag. Squeezing makes it cloudy, so proceed accordingly. When you have the extract, instead of cooking or adding pectin, stir one tablespoon of cornstarch per cup of extract in two tablespoons of water and add it to your extract. Heat this and stir until the desired consistency is reached. It will "jell" to a thicker consistency when refrigerated. You can then freeze it or simply keep it in the refrigerator. Most jelly made in this fashion (especially if you have sweetened your extract with honey or sugar) will keep indefinitely in the refrigerator. For a firmer jell, add a little more cornstarch; for a sauce, add a little less. You can also use commercial pectin or unflavored gelatin, following instructions on the package.

Winnowing. The Indians separated grain from chaff by the following method. The whole grains (such as wild barley or oat) are flamed in a wire basket to remove the coarse chaff; the rest of the grain is trampled or rubbed to remove remaining chaff, and the entire mass is then sifted through the fingers over a cloth in a gentle wind. The wind blows the lighter chaff away, while the heavier grain falls to the cloth. I have also found that after firing the larger chaff off within the wire basket, the entire parched grain and remaining chaff can be ground in a hand gristmill. The grain pulverizes, while the chaff comes through fairly whole. Sifting with a standard flour sifter then removes the chaff from the grain flour. A tea strainer with a fine mesh may be used for this final sifting.

Grinding Wild Flour. For the wild-food enthusiast who does not wish to invest in a gristmill for pulverizing grains and hard fruits, there are other methods. A small hand gristmill may be purchased for about $25, but a meat grinder or a home blender may suffice. To obtain a reasonable facsimile of flour from oat, barley or Carob, for example, even inexpensive blenders will do a good job if the grain or dried fruit is dropped on top of the spinning blades. The rapidity with which the blades spin keeps the product being ground from falling below them so that it repeatedly is spun from the blades, falls back, and is pulverized bit by bit. To keep the product from flying out of the blender upon contact with the blades, the lid must be placed on the blender at the same instant that the material is introduced. The trick is to start the blender and quickly add the material in small amounts. (I place it on the inside of the inverted lid, then quickly place the lid on the blender container, thus emptying the contents of the lid into the container and down onto the spinning blades in one motion.) A less demanding way to accomplish this is to make an inverted funnel with aluminum foil, tape it to the top of the blender container, and add the grain or fruit through the small opening in the top, again making sure the blades are spinning before the material is added.

If a wet product is not an inconvenience, the material to be ground can be added to water and pulverized in the normal blender fashion. There are also some blenders on the market that will pulverize dry materials without the addition of liquid, in which case the above instructions would not apply.

FOOTHILLS AND MOUNTAINS

1. California BAY
   (also California Laurel, Bay Laurel, Oregon
   Myrtle, Pepperwood) *Umbellularia californica*
   (fig. 1) Family: Lauraceae

Description: An evergreen tree with a broad crown. Height
ranges from erect shrub in dry areas to 30 m. Leaves 3-8 cm
long, giving off a spicy aroma when crushed. Yellow-green
flowers in clusters of 6-10. Fruit is a thin-shelled drupe
turning from green to dark purple. Blooms December
through May.

Distribution and Habitat: Common in wooded canyons
and valleys of Chaparral, Yellow Pine Forest, Foothill
Woodland, and Mixed Evergreen Forest, below 5000 ft (1500
m), in coastal mountain ranges and Sierra Nevada from San
Diego Co. to Oregon, where it reaches its maximum size.

Uses—Past and Present: Early Spanish settlers pulverized
the leaves for a condiment, while Cahuillas and other
California Indians used them medicinally. To cure headache,
a piece of leaf was placed inside the nostril or several leaves
were bound on the forehead. A tea made from the leaves was
drunk for headache and stomach ailments. A hot bath using
Bay leaves was used as a treatment for rheumatism. White
settlers adapted this by combining the oil from the leaves with
lard and rubbing it on the body. Mendocino Co. Indians used
the aromatic leaves to repel fleas; they burned boughs to
fumigate lodgings against colds. Even today, small branches
are sometimes placed in chicken coops to prevent lice.

Indians parched or roasted the nuts in ashes, then cracked
and ate them. Today the leaves are prized for a flavor additive
to soups, stews, and meat dishes. They are packaged
commercially in San Francisco and are taking the place of the
European Bay leaf, which they resemble in taste and
appearance, although the California Bay has a much stronger

19

Fig. 1. California BAY *(Umbellularia californica).* 1/2 X.*

taste than its European relative, so a smaller amount is needed for flavoring. The wood is light and fine-grained and can be worked into beautiful bowls; it is usually called Oregon Myrtle and is in such demand for furniture that consideration should be given to the protection of this species.

BAY STEWED TOMATOES

6 whole tomatoes	1 small onion, chopped
dash of salt	1 bay leaf
3 1/2 cups of water	1/8 tsp pepper
2 Tbsp butter or margarine	1/2 tsp Worcestershire sauce
2 Tbsp minced parsley	salt to taste

Combine tomatoes with dash of salt and water. Cover and bring to a boil. Simmer for 15 minutes. Drain, reserving juice.

---

*Magnification of specimen.

Put juice from tomatoes, butter or margarine, parsley, onion, and bay leaf into pan. Cook down to half. Add tomatoes, pepper, Worcestershire sauce, and salt to taste. Serves 6. A 2-lb can of tomatoes may be used instead of fresh tomatoes. Omit the initial simmering of the tomatoes and begin with the drained juice.

An additional recipe can be found under Pickleweed.

2. Sierra BILBERRY
(also Dwarf Bilberry, Dwarf Huckleberry) *Vaccinium nivictum* (fig. 2) Family: Ericaceae

Description: A low, creeping shrub, 5-10 cm high. Leaves glaucous, roundish, 1-2 cm long with gland-tipped teeth. Pink or white bell-shaped flowers occur singly in the axils of the leaves and are 5-6 mm long. Fruit is a round berry, blue-black with a whitish bloom, 5-7 mm in diameter, with a sweet taste. Blooms June through July and bears fruit in August and September.

Distribution and Habitat: Wet meadows and tufts around boulders, 7000-12,000 ft (2100-3700 m) in Subalpine Forest and above the timberline in Sierra Nevada from Tulare Co. north to Mt. Shasta. In the fall, mountain meadows in high Sierra turn a beautiful reddish color as the bilberries and blueberries assume their autumn colors.

Uses—Past and Present: Indians gathered the berries, dried them in the sun or over a campfire, then stored them for use later, much as we do raisins. Pounded with meat and fat, they were sometimes made into pemmican. Though the plants are common, the berries are somewhat scattered. Of excellent flavor, they can be used fresh, cooked, or dried and are considered a delicacy by backpackers. Bilberry fruit and leaves are eaten by the Blue Grouse, the Hermit Thrush, and the Golden-mantled Ground Squirrel. Pika and Raccoon eat the fruit, twigs, and foliage. *V. uliginosum,* the Bog Bilbrry, and *V. coccinium,* the Siskiyou Mountain Bilberry, of northern California are on the rare and endangered plant list of the California Native Plant Society and should be used as emergency food only.

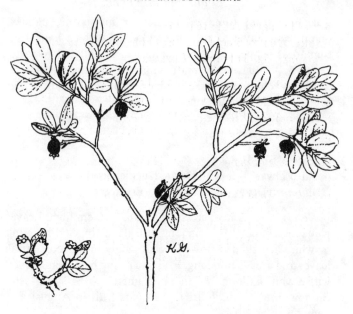

Fig. 2. Sierra BILBERRY *(Vaccinium nivictum)*. Actual size.

OLD-FASHIONED BILBERRY MUFFINS

1 egg, beaten
3/4 cup brown sugar
1/2 cup milk
1 Tbsp melted butter
  or margarine
1 1/2 cups flour

1/4 tsp salt
1 Tbsp baking powder
1 1/2 cups bilberries
  (or blueberries)
2 Tbsp flour

Preheat oven to 425° (hot). Combine egg with brown sugar and milk. Stir in melted butter or margarine. Sift together 1 1/2 cups flour, salt, and baking powder. Quickly combine wet and dry ingredients. Toss bilberries (or blueberries) with 2 Tbsp flour and fold into batter. Fill well-greased and floured muffin cups 2/3 full and bake 12 minutes. Makes 12.

Other recipes may be found under Blackberry, Blueberry, and Currant.

3. California BLACKBERRY
   (also Pacific Blackberry, Bramble Bush) *Rubus ursinus* (fig. 3) Family: Rosaceae

Description: Thorny trailing shrub forming typical briar thickets. Leaves tri-foliate with ovate, dentate leaflets 5-12 cm long. White flowers, in clusters of 2-15. Sepals 5; petals 5; stamens many. Fruit 2 cm long, black, produced summer and fall.

Distribution and Habitat: Damp places, waste places, fields, and canyons below 3000 ft (900 m) throughout most of the state. A similar species, *R. vitifolis,* has bright green leaves, tri-foliate or lobed. Leaflets oval, the terminal to 10 cm long. Inhabits damp places, stream banks, and woods from 4000 ft (1200 m) to beaches, Mendocino to San Luis Obispo counties. *R. procerus,* the delicious Himalaya Berry, with 3-5 leaflets, is widely found from northern California to Washington and is the berry most used in blackberry pie. *R. laciniatus,* with deeply incised leaves, is also common and occurs often with Himalaya Berry. They are equally delicious.

Uses—Past and Present: Indians dried the berries for preservation, then soaked them in water when ready to be used. Fresh berries were pounded to form cakes or mixed with dried meat and fat to make pemmican. A tea made from blackberry roots was the most frequently used remedy for diarrhea among the Indians of northern California. In the eastern United States an entire village of Oneida Indians once cured themselves of dysentery with this tea while neighboring white settlers succumbed to the disease. The root bark was entered into the U.S. pharmacopoeia from 1820 to 1916. Besides having medicinal value, the half-ripe berries were often soaked in water to make a pleasant drink. The Cahuillas of southern California were known to use the berries in this way.

Today the blackberry is popular among campers and hikers, and the leaves (fresh or dried) can be brewed to a fine tea. The young shoots can be boiled and eaten but must be

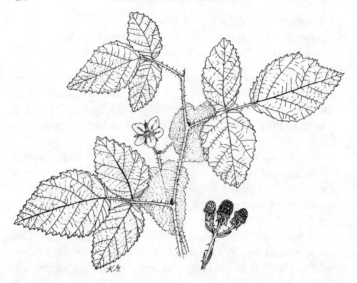

Fig. 3. California BLACKBERRY *(Rubus ursinus)*. 1/2 X.

picked before the thorns harden. The many species of *Rubus* rank at the top of the list of summer foods for wildlife. Many animals and birds depend on the abundant plants; the fruit is eaten by the Band-tailed Pigeon, Western Bluebird, Black-headed Grosbeak, Scrub and Steller's Jay, Mockingbird, Bullock's Oriole, Fox, Song and White-Crowned Sparrow, Western Tanager, California Thrasher, Brown Towhee, Wren-tit, and Portola Wood Rat. The stems and foliage are browsed by the California Mule Deer. Many of these same animals use the dense berry thickets for nesting sites and protective cover.

To avoid thorns, gather blackberries while wearing gloves that have the fingertips cut out.

BLACKBERRY PAN SHORTCAKE

2 1/3 cups biscuit mix   1/2 cup milk
1/4 cup sugar   3 cups blackberries
1/4 cup melted butter   sweetened and
  or margarine   partly crushed

Preheat oven to 425° (hot). Combine biscuit mix, sugar,

butter or margarine, and milk. Bake in an ungreased pan, 15-20 minutes or until slightly brown on top. While still warm, cut into 6 squares and cover each with blackberries. Top with cream if desired. This same recipe can be easily made over a campfire with no butter or margarine and only powdered milk by simmering the berries in a covered pan with a little water, then spooning the shortcake on top and replacing the cover until the shortcake is steamed to a fluffy dumpling. Serves 6.

Thimbleberry, Raspberry, Blueberry, and Currant recipes may be made with blackberries.

4. Western BLUEBERRY
   (also Western Huckleberry) *Vaccinium occidentale* (pl. 1a) Family: Ericaceae

Description: Small shrub 3-7 dm high with thin, rounded leaves 1-2 cm long. Bell-shaped flowers mostly solitary or in small clusters. Petals white or pink, 4 mm long. Berry blue-black with a white bloom. Berry is oval, 6 mm long, sweet when fully ripe. Blooms June through July, fruits in summer.

Distribution and Habitat: Wet meadows and stream sides of Lodgepole and subalpine forests at 5000-11,000 ft (1500-3400 m), Sierra Nevada from Tulare Co. north to Modoc and Siskiyou counties, west to Trinity Summit and British Columbia. One of nine species of *Vaccinium* occurring in varied habitats throughout the state, all of which are edible though not of equal palatability.

Uses—Past and Present: Blueberries and their close relatives, the huckleberries, were relished by the Indians and settlers. Most often they were dried for preservation and used both raw and cooked. Blueberries have appreciable amounts of calcium, phosphorous, and vitamin E. Today they are a favorite fruit to add to pancakes or muffins. Though scattered abundance makes some species difficult to collect, they remain a favorite of campers and backpackers, who frequently place a sheet or canvas underneath the bushes and shake them vigorously to release the berries. For animal uses, see bilberry.

Blueberries can be used in dumplings or shortcake in the

same way as blackberries, and for a recipe for muffins, see bilberry. Chokecherry and currant recipes may also be used.

BLUEBERRY PANCAKES

1 1/2 cups flour	2 Tbsp oil
1 Tbsp baking powder	3 Tbsp sugar or honey
1/2 tsp salt	(optional)
1 cup milk	3/4 cup blueberries,
1 egg	tossed in 2 Tbsp flour

Mix flour, baking powder, salt, milk, egg, oil, and sugar or honey until smooth. Add floured berries to batter at the last minute. Spoon onto hot griddle. (Or use your favorite pancake batter and add 2 Tbsp floured berries for each serving.) Serves 4.

BLUEBERRY PIE

2 qt blueberries	2 Tbsp lemon juice
1/4 cup flour	2 cups brown sugar
2 Tbsp butter or margarine	1 cup whipped cream or
1/2 tsp salt	sour cream (optional)
	1 baked 10-in pie shell

Simmer together 1 qt blueberries, flour, butter or margarine, salt, lemon juice, and brown sugar until mixture thickens. Add another qt of blueberries and pour into baked pie shell. Top with sweetened whipped cream or sour cream, if desired. For variation, try adding 1/2 tsp of nutmeg or allspice to pie filling.

5. BRACKEN FERN

(also Brake) *Pteridium aquilinum* (pl. 1b) Family: Pteridaceae

Description: Fern with erect fronds 3-15 dm tall. Lower portion of fronds three times pinnate with linear segments. Fronds 3-9 dm broad, cut into 3 widely spreading branches, each of which is again subdivided into sessile leaflets. The sori, or fruiting dots, are continuous on the underside leaflet margins. Variable.

Distribution and Habitat: In moist places at low elevations and a common ground cover in forests at high elevations to 10,000 ft (3100 m) throughout the state. It is the most abundant fern in the Sierra.

Uses—Past and Present: The many uses of Bracken Fern extend beyond our continent. Ancient herbalists used the roots and whole plant in drinks for the "spleen" and other internal disorders. The young fronds, called fiddleheads, are used in brewing a beer in Siberia and Norway. The Maoris of New Zealand use the fronds as food. Probably the most intense use of the fern today is in Japan, where the fiddlehead is collected, boiled, and eaten as a vegetable or dried in the sun for storage. The rootstalks produce a form of starch known as *warabi* which is much desired for confections.

Some Indian hunters would feed solely off fiddleheads so their scent would not scare the deer, who fed on the same plant. In the early days of California, miners gathered the young fronds, soaked them for twenty-four hours in water and wood ashes, boiled them, and ate them like asparagus.

Most literature calls for the removal of the tough basal hair and curled top of the fiddlehead before boiling, but I have enjoyed the unprocessed young tips both raw and cooked. They are particularly nice in stews and soups. The uncurled or fully developed fronds should not be eaten; they are reported to poison livestock, although they are an important deer browse. The Bracken Fern is reportedly eaten by the Mountain Beaver.

COMPANY FIDDLEHEADS

4 cups fiddleheads	1/4 cup melted butter or
lemon juice	margarine
salt and pepper to taste	1/2 cup bread crumbs

Pick fiddleheads between 6 and 8 in tall. If reddish hairs are present, brush them off. Steam until tender (or boil). Squeeze a little lemon juice over them, and season with salt and pepper. Combine melted butter or margarine with bread crumbs and pour over the cooked fern. Serves 4.

6. BRODIAEA
(also Blue Dicks, Wild Hyacinth, Ookow) *Dichel-ostemma pulchella* formerly *Brodiaea pulchella* (fig. 4) Family: Liliaceae

Description: A violet-purple (rarely white) flowered relative of the lily, growing from a deep-seated bulb. The funnel-shaped flowers form a compact cluster on top of a single stiff stem 3-6 dm high. Leaves, which usually wither before the flower appears, are 1.5-4 dm long, 5-12 mm wide. Blooms March through May.

Distribution and Habitat: Common on plains and hillsides below Yellow Pine Forest in most of California west of the Sierra Nevada. Uncommon on drier hills of eastern Sierra. A pale blue variety occurs in open desert areas of southern California. Two common *Brodiaeas* that have similar uses are *B. lutea,* the Golden Brodiaea, which has golden flowers, and *B. elegans,* the Harvest Brodiaea, which has deep purple flowers. These species vary in palatability.

Uses—Past and Present: The bulbs (technically corms) of the Brodiaea are thought by some to have been one of the most important underground food plants of the Sierra Indians, such as the Miwoks. The bulbs were gathered in large quantities with digging sticks and great feasts were held. They were eaten raw, fried, boiled, and roasted. Some have such a sweet, nutty flavor that no cooking is required, while others are best fried or roasted. Many are considered tastier than potatoes, and I have used the flowers in wild salads. When gathering Brodiaea in mountain meadows, pick only bulbs with flowers, since the white-flowered Death-Camas bulbs may grow in the same habitat (see pl. 8f). Brodiaea bulbs are found from 1.5 to 2 dm below the ground and are up to 2.5 cm in diameter. The bulbs are eaten by the Botta Pocket Gopher of the San Joaquin Valley. Brodiaeas are beautiful wild flowers and should be sampled only in emergencies or when they are growing in great abundance.

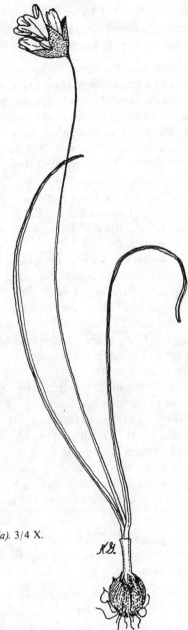

Fig. 4. BRODIAEA
(Dichelostemma pulchella). 3/4 X.

7. Leichtlin's CAMAS and Common CAMAS
   (also Quamash) *Camassia leichtlinii* and *C. quamash* (fig. 5) Family: Liliaceae

Description: Perennial herb, about 6 dm in height. Bulb resembles a small onion, without the smell, frequently with a blackish coat on the outside. Leaves in basal whorl, grasslike. Flowers are showy, blue or blue-violet with petals 2-4 cm long. Blooms May through August.

Distribution and Habitat: Wet meadows in mountains to 8000 ft (2500 m) in Montane Forest and Mixed Evergreen Forest of coastal mountain ranges from Marin Co. north, and Sierra Nevada from Mono and Tulare counties north. One variety of camas has smaller petals and is found in the Sierra Nevada from El Dorado Co. north to Siskiyou and Modoc counties.

Uses—Past and Present: Because of the relatively large size and good flavor of the bulb, this was a very important Indian food plant. Wars were fought over collecting rights. Bulbs were sometimes boiled down to make a syrup, but more often they were baked in pits. They were often dried for future use or pounded into cakes which could be dried. White settlers were fond of camas pie, but some foragers today object to the mucilaginous quality.

As in the use of all unfamiliar foods, camas should not be eaten in excess or some emetic effects may occur. Extreme care should be taken to insure that the flower is attached to the bulb, for the Death-Camas, which has white or cream-colored flowers, may grow alongside this plant (see pl. 8f). Camas bulbs may be used in Squaw Root recipes and in soups and stews. However, this is a beautiful wildflower that you will normally want to observe rather than eat.

8. Holly-leaved CHERRY
   (also Islay) *Prunus ilicifolia* (fig. 6) Family: Rosaceae

Description: Shrub or small tree 1-8 m tall, with gray or reddish-brown twigs. Irregularly curled leaves are oval, 2-5 cm long, with sharp teeth on edges. Flowers are white, 2-3

Fig. 5. Leichtlin's CAMAS
(Camassia leichtlinii). 3/4 X.

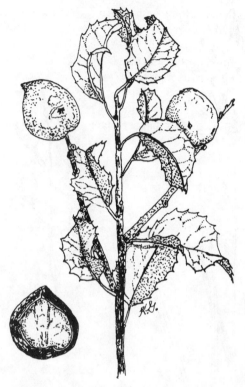

Fig. 6. Holly-leaved CHERRY *(Prunus ilicifolia)*. Actual size.

mm long. Fruit ranges from red to yellow, 12-15 mm long, with thin, sweetish pulp. Blooms April through May with fruit ripening in summer and fall.

Distribution and Habitat: Common on dry slopes and fans below 5000 ft (1500 m), especially in Chaparral, from Baja California through coastal mountain ranges to Napa Co. Extensively planted as an ornamental.

Uses—Past and Present: The Cahuillas favored this cherry, which was abundant in the San Jacinto Mountains. They used the kernel primarily, extracting it from the pit, crushing it in a mortar, leaching it in a sand depression (see Notes on Plant Preparation Techniques), and boiling it into a soup. The fruit was also pressed to make a drink, and the bark was used to make a tea for treating colds. The cherries were often gathered along with acorns, which are often found nearby. Hikers enjoy nibbling on the thin, sweet pulp. The kernel is dried, crushed, and leached by nut fanciers. The leached nut may be substituted into acorn recipes, and a sweet sauce may be made with the thin pulp by simmering it in a little water and adding lemon and sugar to taste. The glossy foliage of Islay is frequently used in making Christmas decorations. For animal uses, refer to Chokecherry.

9. Sierra CHINQUAPIN
(also Bush Chinquapin) *Chrysolepis sempervirens* formerly *Castanopsis sempervirens* (pl. 1c) Family: Fagaceae

Description: Low-spreading, round-topped bush 0.5-2.5 m high, with smooth brown or gray bark. Leaves oblong, 3-7.5 cm long, yellowish gray-green above, golden or rusty beneath. Fruit a burr 2-3 cm long, opening with 4 valves having 1-3 nuts. Blooms July through August.

Distribution and Habitat: Dry rocky slopes and ridges in Montane Forest from 2500-11,000 ft (800-3400 m). San Jacinto Mountains to San Gabriel Mountains, Sierra Nevada from Kern Co. north, coastal mountain ranges from Lake Co. north. *Chrysolepis chrysophylla,* the Giant Chinquapin, is a tree 15-45 m tall, occurring on slopes below 1500 ft (500 m) near coast from Mendocino to Del Norte counties, and may be used in the same way. A variety of this species, known as the Golden Chinquapin, is a shrub occurring on gravelly or rocky ridges and slopes 1000-6000 ft (300-1800 m) in coastal mountain ranges from Santa Lucia Mountains to Del Norte, Trinity, and Siskiyou counties.

Uses—Past and Present: The Cherokee Indians of the southeastern United States steeped the leaves of the eastern Chinquapin and employed the tea as an external wash for fevers due to colds and flu. This liquid has also been used in American medicine as an astringent. A relative to the domesticated chestnut, the nuts of the chinquapin are sweet and are edible raw or roasted. Rodents, especially squirrels, prize the nuts, which are gathered early in the fall when the burrs begin to open.

CHINQUAPIN SHORTBREAD

1 cup butter or margarine   pinch of salt
3/4 cup brown sugar   1 cup chopped
2 cups flour   chinquapin nuts
1/2 cup cornstarch

Preheat oven to 325° (slow mod.). Cream butter or margarine with brown sugar. Blend in flour, cornstarch, and salt. Knead with your hands until smooth, then add chinquapin nuts. Pat onto ungreased cookie sheet to a thickness of about 1/4 in. Pierce frequently with a fork and bake for 35 minutes. Cut while still warm. Serves 12.

10. Western CHOKECHERRY

Prunus virginiana var. *demissa* (pl. 1d) Family: Rosaceae

Description: Deciduous shrub or small tree 1-5 m high, with smooth gray-brown bark. Wide green leaves are oblong, 3-8 cm long, finely toothed. Flowers in clusters, 5-10 cm long at ends of short branches, with white petals. Fruit is a small cherry hanging in strings, round, 5-6 mm thick, dark red or black, and palatable when fully ripe. Blooms May through June, fruits in summer and fall.

Distribution and Habitat: Damp places in woods and on brushy slopes and flats near streams below 8200 ft (2500 m). Mostly in the mountains from San Diego Co. north through coastal mountain ranges and Sierra Nevada to Washington. Occurring in Chaparral, Foothill Woodland, and Yellow Pine Forest. The Western Chokecherry is said by some authorities to be the most widely distributed tree in North America, probably due to the birds' love for its fruit and subsequent scattering of the seeds.

Uses—Past and Present: Indians and pioneer settlers used the chokecherry for syrup, soups, wine, and other fruit concoctions. Occasionally, the cherries were ground up (stones and all) and dried into cakes which were later soaked in water, mixed with flour and sugar, and used to make a sauce. Pemmican was also made from the cherries. Tea made from the inner tree bark was used to check diarrhea and nervousness. Although the flavor varies from plant to plant, it is the general opinion that the fresh fruit is not as desirable as the dried or cooked fruit. Many people prefer to mix the chokecherry half-and-half with apple for a fine jelly; however, I find the pure product quite delectable when made with fully ripe cherries. The seeds and leaves contain hydrocyanic acid, which has poisoned livestock, but boiling removes this acidity. The wood can be used for bows and arrows.

Many animals eat the chokecherry: the Ring-necked Pheasant prefers the fruit and the buds; Scrub and Steller's Jays, the Phainopepla and Olive-backed Thrush eat the fruit; Raccoons like the fruit, bark, and wood; various species of chipmunk and the Large-eared Wood Rat prefer the fruit alone; and Mule Deer eat the twigs and foliage.

SPICED CHOKECHERRY JAM

3 cups chokecherries, pitted
2 1/2 cups water
juice and pulp of 3 oranges or lemons
1 1/2 tsp ground cloves
1 1/2 cups honey or 2 1/4 cups sugar
liquid pectin

Bring to a boil the chokecherries, water, juice and pulp of oranges or lemons, cloves, and honey or sugar. Simmer for 10 minutes until sauce is clear and sugar is dissolved. Adjust taste with orange or lemon or with sugar. To every 3 cups of sauce, add one 6 oz. bottle of liquid pectin. Boil hard for 1 minute and pour into sterilized jars. Seal with melted paraffin and cover with waxed paper or use lids. Makes 8 cups. This excellent jam makes beautiful gifts.

CHOKECHERRY SYRUP

3 cups chokecherries, unpitted
1 1/2 cups water
1 1/2 cups honey or 2 cups sugar
juice of 3 lemons
2 Tbsp cornstarch
3 Tbsp water

Crush chokecherries and add water, honey or sugar, and lemon juice. Simmer for 15 minutes and add more sugar if desired. Run through a cheesecloth or jelly bag, squeezing a little. Return juice to saucepan and cook until thickened or add 2 Tbsp cornstarch dissolved in 3 Tbsp water to each 3 cups of sweetened juice. Simmer until desired consistency is reached. Pour into sterilized jars and seal. Makes 4 cups. This syrup may be used in jellies and on pancakes or ice cream.

11. Squaw CURRANT
Ribes cereum (pl. 1e) Family: Saxifragaceae

Description: Erect shrub 1-12 dm tall. Leaves 1-4 cm wide, glandular, usually 3-5 shallow lobes, which are rounded with minutely scalloped edges. Flowers in clusters of 3-7. Each flower is 6-8 mm long. One variety has 5 to 10 flowers per cluster and shiny leaves. Berry round, red, 6 mm in diameter, with several small seeds. Blooms June through July, with fruit appearing in summer.

Distribution and Habitat: Dry slopes, 5000-12,000 ft (1500-3800 m), Pinyon-Juniper Woodland to Alpine Fell-fields from Santa Rosa and San Jacinto ranges north through the Sierra Nevada to Siskiyou and Modoc counties and British Columbia. There are thirty species of currant and their close relatives, the gooseberries, recognized in California. All are considered edible, though not always palatable. They are easily recognized by their characteristic leaf shape, which resembles a maple leaf with rounded points, their fleshy fruit with several seeds, and their dried flower, which is often still attached to one end. Some of the more common and delicious species are: *R. montigenum,* the Alpine Prickly Currant; *R. malvaceum,* the Chaparral Currant; *R. nevadense,* the Sierra Currant; and *R. aureum,* the Golden Currant. Other species are listed under Goose-berry.

Uses—Past and Present: Pemmican, that staple of the Indian diet, was made from dried currants by pounding them with meat and fat. This was formed into loaves or cakes and used when traveling. Lewis and Clark feasted on a great variety of wild currants that they claimed were better than their native cultivated forms. Squaw Currant was often used as the chief ingredient in treating stomachache, while young leaves and twigs of several species were boiled as a vegetable in times of need. Today currants are so abundant they still can be gathered in quantity. They are high in vitamin C, phosphorus, and iron. Currants are a favorite fruit for jellies, jams, sauces, and pies.

Shaking the bushes over sheets of plastic or blankets is a popular method of gathering the berries. Nectar-filled flowers are considered good trail snacks and, at times, sweeter than the fruit. Currants have long been the object of extermination programs by the Forest Service because they are intermediate hosts to a rust which attacks five-needled pine trees. The wood makes good arrow shafts. Though the fruits still have much value for songbirds, chipmunks, ground squirrels, and other animals, the importance of currants and gooseberries was probably greater in the past.

OLD-FASHIONED OSGOOD PIE

1 Tbsp melted butter	1/4 tsp nutmeg
or margarine	1/2 cup chopped walnuts
3/4 cup brown sugar	1/2 cup currants
1 Tbsp vinegar	(fresh or dried),
2 eggs, separated	tossed in 3 Tbsp flour
1/4 tsp cinnamon	1 9-in pie shell

Preheat oven to 350° (mod.). Combine melted butter or margarine, brown sugar, vinegar, and egg yolks. Add cinnamon, nutmeg, and walnuts. If using dried currants; soak in hot water for 5 minutes, then add floured currants to rest of ingredients. Beat egg whites until stiff but not dry and fold into the mixture. Turn into pie shell and bake for 45 minutes. Serve warm, with a dollop of whipped cream if desired. Serves 4-6.

ORANGE-CURRANT BREAD

2 large oranges	1 egg
3 Tbsp melted butter	2 cups flour
or margarine	1 cup sugar

1 tsp baking powder	1 cup currants
1/2 tsp. baking soda	(fresh or frozen)
1/2 tsp salt	1/2 cup chopped walnuts
	(optional)

Preheat oven to 375° (quick mod.). Blend the orange juice and grated rind with melted butter or margarine and egg. Sift together flour, sugar, baking powder, baking soda, and salt and add to liquid ingredients. Blend in currants and chopped walnuts. Bake in a greased loaf pan for 1 hour. Serve warm or cold with butter and honey. Serves 6.

These recipes are for sweet red or orange currants. Tart currants or gooseberries need less lemon and more sugar. Other recipes which are adaptable for currants may be found under Bilberry, Blueberry, Cherry, and Blackberry.

12. Miner's DOGWOOD,
Cornus sessilis (pl. 1f) Family: Cornaceae

Description: Shrub or small tree with leaves 4.5-9 cm long, 2-3.5 cm broad, oval with light hair on the veins underneath. Petioles with 2 pairs of deciduous bracts at base that are often brown with yellow edges and 1 cm long. Small white flowers occur in umbels from March through April. Fruit is 1-1.5 cm long, turning from white to dark red, with one large seed maturing in summer.

Distribution and Habitat: Stream banks from 500-5000 ft (150-1500 m) in Redwood Forest and Yellow Pine Forest. Coast mountain ranges from Tehama, Trinity, and Humboldt counties to Siskiyou Co. and in the Sierra Nevada from Calaveras Co. north. There are five other species of Dogwood spanning the length of California, all of which have edible berries. Some are bitter, while those of *C. canadensis,* the Bunchberry, seem somewhat insipid. Miner's Dogwood has sweet berries when fully mature, and the berries which have fallen from trees and partially dried in the sun are especially good.

Uses—Past and Present: Dogwood is an old Indian remedy for fever. The dried root or bark was boiled for a tea. Early explorers used this tea in place of quinine when that medicine was in short supply. It has also been reported that the peeled twigs were used as toothbrushes for their supposed whitening effect on teeth. Other tribes used the bark as an astringent or for smoking.

Today the fruit of the Miner's Dogwood can be the most useful and enjoyable product of all the dogwoods. It can be picked when fully mature or already partially dried as mentioned above. Seasoning with a little lemon, spices, and sugar makes a superior syrup.

Animals enjoy the dogwood also. The fruit is eaten by Band-tailed Pigeon, Crow, Mockingbird, Bullock's Oriole, Russet-backed Thrush, and Cedar Waxwing; the twigs and foliage are eaten by the Mule Deer and Mountain Beaver.

CAMPFIRE CRÊPE SAUCE

1 cup dogwood berries
  (dried)
1/2 cup water
juice of 1/4 lemon

sugar to taste
  (approx. 3/4 cup)
1/2 cup red wine

Combine all ingredients and simmer until syrupy. May be served over thin pancakes. The alcohol is dissipated in the cooking.

13. Desert ELDERBERRY
(also Elderberry, Southwestern Elderberry) *Sambucus mexicana* (cover photo, upper left) Family: Caprifoliaceae

Description: Deciduous shrub or small tree with compound, opposite leaves. Has 3-5 leaflets, mostly 1.5-6 cm long. Numerous small cream-colored, pleasantly scented flowers occur in umbels 3-10 cm across. The twigs and leaves have a rank smell. The BB-sized berries are blue with a whitish bloom. Blooms March through September with fruit following if water is available.

Distribution and Habitat: Open flats and valleys as well as canyon slopes below 4500 ft (1400 m), many plant communities from Baja California to Lake and Glenn counties. *S. caerulea,* Blue Elderberry, has a similar appearance but has more leaflets. The Black Elderberry, *S. melanocarpa,* has black berries without the bloom and occurs in the eastern Sierra and Siskiyou Co. *S. mexicana* is often used as an ornamental. In late summer and early fall the branches of the elderberry are often bent with their heavy load of fruit. I have found the mountain plants to yield the juiciest and best-flavored berries, although the well-watered lowland *S. mexicana* pictured on the cover is also an excellent plant.

Uses—Past and Present: Foragers and herbalists both ancient and modern have prized the elderberry. Many Indian tribes across the nation, including the Mohegans of New Jersey, Menominees of Michigan, and California's Cahuillas, brewed the blossoms into a tea which was used for fevers, upset stomachs, colds, and flu. It was also considered good for newborn babies and for teeth. A wash of boiled flowers or leaves was used to treat wounds. White settlers used these native remedies as well. Elderberry was entered in the U.S. pharmacopoeia for many years prior to 1905 and has been analyzed as one of nature's richest sources of vitamin C.

The stems were used to make arrow shafts, whistles, and flutes. The musical instruments made from the hollowed-out stems gave it the name Tree of Music. A dye from both the twigs and fruit was used in basketry. Some historians believe that the cross of Christ was made from the Elder Tree.

Modern foragers have also learned from the Indians that both the flowers and fruit are edible. Present-day adaptation of fat-fried Indian fritters made from the flowers is to dip them in batter, deep fry, and sprinkle with sugar. If care is taken to use just the smallest stems attached to the umbels, the fried blossoms are a gourmet treat. The flowers can be shaken from the stems to add flavor and fresh vitamins to pancakes, muffins, and cakes. Although the fresh fruit is not always palatable, after drying or cooking it can be made into delicious sauces, jellies, wines, and syrups with a flavor similar to Boysenberry. Wine is also made from the flowers.

Although the fresh fruit makes some people nauseated, dried or cooked elderberries produce no such effects.

Because of the delicious taste of products and the ease of gathering large quantities, the elderberry is a favorite in the households of many foragers. Please note that the red-berried elders of the north coast are inedible.

The elderberry is an important source of summer food for many songbirds such as the Western Bluebird, House Finch, Red-Shafted Flicker, Ash-throated Flycatcher, Black-headed Grosbeak, Scrub and Steller's Jays, Ruby-crowned Kinglet, Bullock's and Hooded Oriole, and Phainopepla.

BASIC ELDERBERRY SYRUP

1 qt elderberries	juice of 1 lemon
3 cups water	1 Tbsp cornstarch
1/4 cup sugar or honey	or flour

Crush elderberries, add 1 cup of water and sugar or honey, and simmer for 15 minutes. Strain, then add 2 cups of water to the seeds and pulp and strain again. Add to the liquid the lemon juice and adjust sugar if desired. Bring to a boil and thicken slightly by stirring one Tablespoon cornstach or flour in one Tablespoon cold water and stirring this into the simmering syrup. Makes 5 cups. This syrup has few equals when used over pancakes or ice cream.

ELDERBERRY CREAM PIE

3 eggs, separated	2 Tbsp grated orange
3/4 cup elderberry syrup	or lemon peel
1 envelope	1/4 tsp cream of tartar
unflavored gelatin	1/4 cup sugar
1/3 cup sugar	1 cup heavy cream
pinch of salt	1 baked 9-in pie shell

Blend over heat until smooth the egg yolks, elderberry syrup, unflavored gelatin, 1/3 cup sugar, and salt. Do not boil. Add grated orange or lemon peel and pour into a bowl; refrigerate until firm. Beat the egg whites until stiff and add cream of tartar and 1/4 cup sugar, beating continuously. Beat heavy cream until fluffy and fold half into the egg-white mixture. Fold the egg-white mixture into the refrigerated sauce. Pour

into pie shell and garnish with remaining whipped cream.
Serves 6.

14. Douglas FIR,
   Pseudotsuga menziesii (fig. 7) Family: Pinaceae

Description: Forest tree to 70 m tall with a narrow crown.
Branches slender, crowded, with long pendulous side
branches; bark has dark ridges. Branchlets pubescent for 3-4
years; needles 2-3 cm long, rounded on ends, yellow-green,
sometimes bluish. Leaves appear 2-ranked because of
twisting and are grooved, dark above and paler beneath.
Female cones hang down, have rounded flexible scales with a
light hair on them and distinctive 3-toothed bracts about 6
mm wide. Seeds are about 6 mm long with wings. Cones ripen
in fall.

Distribution and Habitat: Moist slopes usually below 5000
ft (1500 m) in Yellow Pine and North Coastal Forest from
Santa Barbara Co. north in the coast ranges and in the Sierra
Nevada from Fresno Co. north. It is the most important
lumber tree in North America. *P. macrocarpa,* Big Cone
Spruce, occurs on dry slopes and in canyons below 6000 ft
(1800 m) in Chaparral and Yellow Pine Forest from Santa
Barbara Co. to San Diego Co. *P. macrocarpa* is generally
smaller and has wider bracts on the cones and winged seeds
about 12 mm long. The two are easy to differentiate because
of their ranges: the Big Cone Spruce is prevalent in the
southern part of the state while the Douglas Fir is found in
the central and northern parts.

Uses—Past and Present: The Douglas Fir was used
extensively by the California Indians for lumber, harpoon
shafts, and other implements. The pliable roots were used in
weaving baskets. A tea made from the fresh needles was a
favorite drink of the Yuki Indians in the north. In British
Columbia, the young twigs were sometimes boiled to make
tea. As with other cone-bearing trees, the cambium layer was
utilized as an emergency food. Today the trees are often sold
as Christmas trees. The small winged seeds of this conifer are

Fig. 7. Douglas FIR *(Pseudotsuga menziesii).* 1/2 X.

eaten by Western Tree Squirrel, Douglas Chickaree, and the Red Tree Mouse. The needles and some of the male cones are the primary winter food of the Blue Grouse.

15. FIREWEED
(also Willow Herb) *Epilobium angustifolium*
(fig. 8) Family: Onagraceae

Description: Perennial willow-like herb with simple, nonbranching stems. It is 6-25 dm tall with alternate leaves that are 7-20 cm long and lanceolate. Flowers are in long, terminal racemes with lilac-purple or rose petals. The four petals are clawed, 8-18 mm long. There are 8 stamens. The fruit develops below the flower and splits into 4 valves that release numerous tufted seeds. Variable in color and size. It blooms from July through September.

Distribution and Habitat: In disturbed areas and fairly moist places, mostly below 9000 ft (2800 m) in mountains from San Diego Co. to Modoc and Siskiyou counties. Occurs in northern coastal mountain ranges to the seacoast.

Uses—Past and Present: The young shoots, leaves, and flowering stalks are eaten in Europe and Asia as well as North America. The more mature plants can be used for tea or the

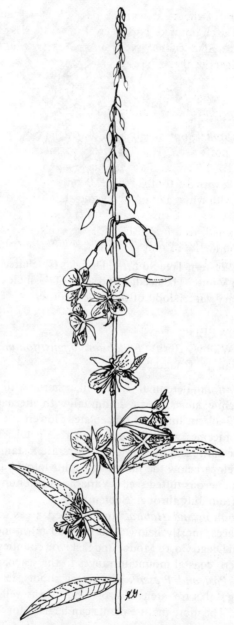

Fig. 8. FIREWEED *(Epilobium angustifolium)*. 3/4 X.

peeled core of the stems can be chewed as a snack. It is browsed by chipmunks and deer. Fireweed is a valuable honey plant. Smaller species of *Epilobium* have been used by some foragers in the same way.

16. Sierra GOOSEBERRY
Ribes roezlii (pl. 2a) Family: Saxifragaceae

Description: Stout shrub 5-12 dm high. Young growth hairy but not bristly. Has 1-3 spines at the nodes. Leaves rounded, 1.2-2.5 cm in diameter, dark green, paler beneath than above, with 3-5 toothed lobes. Petioles 6-18 mm long, 1-2 flowers with white hair on ovary and gland-tipped bristles. Unattached portion of flower tube about 6 mm long, purplish, sepals dull purplish-red, and petals whitish. Berry purple, round, 14-16 mm in diameter, with long pubescent spines. Variable. Fruits in summer.

Distribution and Habitat: Dry slopes mostly 3500-8500 ft (1100-2700 m) in Yellow Pine Forest, Red Fir Forest from San Diego Co. through Sierra to inner coast ranges and Trinity and Modoc counties. Another common species of gooseberry, *R. velutinum,* with a smaller berry, occurs in Sagebrush Scrub, Pinyon-Juniper Woodland and some pine forests from the San Gabriel Mountains to Trinity, Siskiyou and Modoc counties.

Uses—Past and Present: Gooseberries were much used by colonists and Indians alike. Gooseberry pie was a favorite in early times. As was its close relative the currant, the gooseberry was formed into pemmican with meat and fat. It is high in vitamin C and because of its delicate flavor and good size, is a favorite among backpackers. For animal uses, refer to Currant.

GOOSEBERRY PIE

1/4 cup flour
2 Tbsp butter or
  margarine
1/8 tsp salt
1/4 cup lemon juice
1/2 cup sugar

1/4 cup water
2 cups gooseberry pulp
  (Mash berries in a sieve to
  remove hairs and seeds)
1 baked 10-in pie shell

Blend in a saucepan all ingredients except 1 cup of gooseberries; simmer until thickened. Remove from heat and add 1 more cup of gooseberries. Pour into pie shell and chill. (For a double-crusted pie, pour filling into unbaked crust, cover with top crust, pierce several times with a fork, and bake 20 minutes at 350°.) Whole berries may be used in areas where they occur without bristles or too many seeds. Serves 6.

17. Mountain GRAPE
(also Barberry) *Mahonia pinnata* formerly
Berberis pinnata (pl. 2b) Family: Berberidaceae

Description: Evergreen shrub 3-16 dm high. Alternate leaves are compound with 5-9 hollylike leaflets, sharp pointed and glossy green above, paler beneath, 2.5-5 cm long, with bristlelike teeth on each side. Attractive, small yellow flowers occur in elongated terminal clusters followed by blue berries with a gray bloom, 6 mm long. Inner wood is yellow. Blooms March through May, fruits in summer.

Distribution and Habitat: Rocky slopes below 4000 ft (1200 m). Pine and Redwood forests of coastal mountain ranges from San Gabriel Mountains and San Diego Co. to southern Oregon. *B. aquifolium,* Oregon Grape, is taller. *B. haematocarpa* has a white bloom and reddish-purple berries and occurs in dry rocky places of the eastern Mojave Desert, New York Mountains, and Old Dad-Granite Mountains. *B. repens* is a low, creeping form with 5 gray-green leaflets in open shade woods from Inyo Co. north.

Uses—Past and Present: Navajo Indians used the roots of several species to make a yellow dye for their baskets. Other Indian groups boiled the root to drink as treatment for venereal disease or fever. California Indians had the best use of all, perhaps, using this same tea for an aperitif. The state flower of Oregon, the Oregon Grape, was used by the Kwakiutls for a tea to aid digestion, while all tribes, ancient and modern, ate the berries. Berries were apparently eaten fresh and not dried as many others were. Mountain Grape berries were used for making a paint for decorating bows and arrows by the Karoks.

Today the tart berries are a favorite for jelly due to their natural pectin. Mule Deer, Mountain Sheep, elk, and other browsers eat the leaves, despite their prickles. The plant is eaten by birds to a limited degree, although grouse, pheasant, and Cedar Waxwing are fond of the berries. *B. pinnata* ssp. *insularis,* found in the Channel Islands, *B. higginsiae, B. nevinii,* and *B. soneii,* found in various parts of the state, are all on the rare and endangered plant list of the California Native Plant Society.

MOUNTAIN GRAPE JELLY

1 qt Mountain Grape berries	1/2 cup water
	4 cups sugar
1 lemon or	1 bottle liquid pectin
1/4 cup apple juice	cinnamon

To the cleaned, crushed berries add the lemon or apple juice and water, and simmer for 15 minutes. Place in a jelly bag and let the juice drain out. (Do not squeeze.) Reheat juice to boiling. For each 3 1/2 cups of juice, add 4 cups of sugar and 1 bottle of liquid pectin. Bring to a boil as rapidly as possible and boil hard for 1 minute, stirring constantly. Remove from heat, skim off foam, and pour into sterilized jars. A pinch of cinnamon added to the jar and stirred in will give the jelly an extra tang. Because the berries vary in amount of tartness and juice, the lemon and apple juices may be varied according to taste. Seal with paraffin or two-piece lids. (For Instant Jelly, see Notes on Plant Preparation Techniques.) Makes 6 cups.

18. Wild GRAPE

Vitis californica (fig. 9) Family: Vitaceae

Description: This familiar vine needs no introduction. The leaf shape, tendrils, and sweet fruit have the same appearance as cultivated grapes. The fruit is purple with a whitish bloom. The seeds have an acceptable taste. Blooms May through June; fruit ripens in late summer.

Distribution and Habitat: Stream banks and canyons below 4000 ft (1200 m) in Mixed Evergreen Forest, Foothill

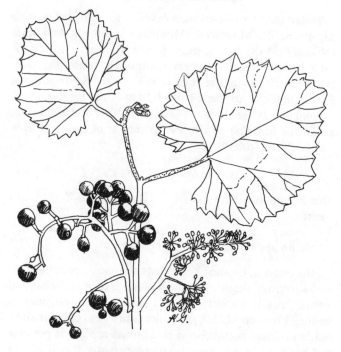

Fig. 9. Wild GRAPE *(Vitis californica).* 3/4 X.

Woodland, and Oak Woodland of the coastal mountain ranges from San Luis Obispo to Siskiyou counties, and the Central Valley and Sierra Nevada foothills from Kern Co. north. *V. girdiana,* the Desert Grape, has black berries with very little or no bloom and occurs from Inyo and Santa Barbara counties south, occasionally on the desert edge.

Uses—Past and Present: The Indians used the Wild Grape much as we do today, utilizing the fresh fruit or drying it in the sun for future use. Arizona's Pueblo Indians are believed to have cultivated it, while the Cherokees in the southern United States boiled geranium root together with Wild Grape and used the liquid to rinse mouths of children affected with thrush. Juice from the leaves was used to treat diarrhea and lust in women. The Cahuillas made wine from the fruit; the Pomo Indians used the vines for weaving baskets. Survival books suggest that you may obtain water in emergencies by cutting the vines off near the top first, then cutting at the bottom and letting them drain into a container or your mouth.

Boiled grape leaves are often used to wrap other foods, and the tendrils make a pleasantly sour snack when raw. The skin of the Wild Grape is too tough for jam and the seeds are large, but jelly can still be obtained from both the pressed ripe or unripe fruit. Grape pie is a favorite of foragers.

WILD GRAPE JELLY

1/3 cup water
1 qt Wild Grape

4 cups sugar
1 bottle liquid pectin
(6 oz.)

Add water to cleaned grapes (half of them underripe), crush and simmer for about 15 minutes. Strain through a cheesecloth or jelly bag. Let the juice sit overnight in a crock or bowl to settle out crystals. For each 4 cups of juice, add 4 cups of sugar and follow the instructions on a liquid pectin bottle or boil the jelly until the juice first drips off the spoon, then comes off in sheets (the "jelly test"), at which point the jelly is done. Makes 6 cups.

For pies, the grapes must be skinned then the pulp sieved to remove the seeds. The recipe for Blueberry Pie can then be followed.

19. California HAZELNUT
(also Filbert, Beaked Hazelnut) *Corylus cornuta*
(fig. 10) Family: Betulaceae

Description: Spreading shrub 2-6 m high with smooth bark. Leaves rounded, 4-7 cm long, doubly toothed, pale beneath with soft pubescence. Bristly husk surrounding the fruiting flower forms a tube 1.5-2.5 cm long. Nut oval, 1.2-1.5 cm long. Flowers January through April with nut following.

Distribution and Habitat: Damp slopes below 7000 ft (2100 m) in many plant communities of northern California from Santa Cruz in coast ranges north, and Tulare Co. north in Sierra.

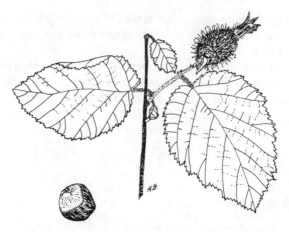

Fig. 10. California HAZELNUT *(Corylus cornuta).* 1/2 X.

Uses—Past and Present: Hazelnuts were eaten to some extent by the Indians of California. In England a cultivated form of hazelnut is planted for numerous purposes. The wood is used for making charcoal from which artists get their drawing pencils. It is tough and flexible and used for javelin shafts and fishing poles. Roots are used for veneering in cabinetry and in refining wine. Shelled nuts easily shed their skins when roasted in a slow oven for 30 minutes, cooled, then rubbed with a towel. Hazelnuts are delicious right off the bush, but may be roasted and used in innumerable recipes. Squirrels, chipmunks, and other rodents feed on the nuts; grouse eat the catkins, rabbits and deer browse the entire plant.

HAZELNUT CRUMB COOKIES

1 1/2 cups flour
1 tsp baking powder
1/2 tsp salt
2 cups brown sugar

5 Tbsp butter
  or margarine
2 eggs, separated
1 tsp vanilla
1 cup chopped hazelnuts
  (or walnuts or acorns)

Preheat oven to 325° (slow mod.). Combine flour, baking

powder, salt, and 1 cup brown sugar. Blend butter or margarine and slightly beaten egg yolks, and add to dry ingredients. Mix until crumbly. Pat into 8x12-in ungreased baking pan. Beat egg whites slowly with 1 cup brown sugar until thoroughly blended. Add vanilla and hazelnuts. Spread on top of crumb mixture. Bake for 1 hour. Cut into squares.

GLACÉED HAZELNUTS

1 cup hazelnuts	1/2 tsp vanilla
1/2 cup sugar	3/4 tsp salt
2 tsp butter	
or margarine	

Combine hazelnuts, sugar, and butter or margarine in a skillet. Cook, stirring, over low heat until nuts are toasted and sugar is golden brown (about 15 minutes). Stir in vanilla. Spread nuts on a sheet of foil and sprinkle with salt. Cool and break into pieces.

20. California HUCKLEBERRY
   (also Evergreen Huckleberry or Shot Huckleberry) *Vaccinium ovatum* (pl. 2c) Family: Ericaceae

Description: Stout shrub 1-2.5 m tall. Oval, leathery leaves are 1.5-4 cm long and have small teeth. Flowers are 5-7 mm long, white to pink, bell-shaped, and occur in small clusters in leaf axils. Berry is oval (or pear-shaped with a bloom), 6-9 mm long; color varies but the berry is black and sweet when ripe. Berry is crowned with withered flower. Blooms March through May, with fruit following.

Distribution and Habitat: Dry forested slopes and canyons below 2500 ft (800 m). Most common on coast from Del Norte to northern Santa Barbara Co., with a few scattered locations as far south as San Diego. *V. membranaceum,* the Mountain Bilberry, has yellow flowers and dark red to black fruit and occurs in moister areas at higher altitudes. *V. parvifolium,* the Red Huckleberry, has greenish or whitish flowers and a bright red berry. It is found in the Sierra from

Del Norte and Siskiyou counties to Fresno. *V. uliginosum* and *V. coccinium* of northern California are on the rare and endangered plant list of the California Native Plant Society.

Uses—Past and Present: Huckleberries have been used much as blueberries. The Iroquois often picked them in snake-infested areas and at such times would smear their moccasins with hog fat to frighten away rattlers. For animal uses, refer to Bilberry.

Huckleberries may be substituted in recipes given for Bilberry, Blueberry, Currant, Elderberry, and Gooseberry.

21. LEMONADE BERRY
(also Coast Sumac) *Rhus integrifolia* (fig. 11)
Family: Anacardiaceae

Description: Rounded shrub, 1-3 m tall, with finely pubescent reddish twigs. Leaves leathery and aromatic, mostly smooth margins but some with shallow teeth. Leaves 2.5-5 cm long, 2-3 cm wide. Small flowers in closely packed clusters are white to rose. Fruit is a sticky, sour drupe with velvety pubescence, reddish in maturity, flattened, about 1 cm in diameter and 7 mm long. Blooms February through May, with fruit following.

Distribution and Habitat: Ocean bluffs and canyons, dry places below 2500 ft (800 m). In Coastal Sage Scrub and Chaparral, Santa Barbara Co. to Baja California and inland to western Riverside Co. *R. ovata,* the Sugar Bush, has leaves folded along the midrib and slightly smaller fruit with the same acid taste. It occurs on dry slopes away from the coast from Santa Barbara Co. to northern Baja California and the western edge of the Colorado Desert.

Uses—Past and Present: These sumac species were often used in making a drink; the berries were soaked in water and the fine hairs strained out. Indians dried the berries for preservation, then soaked them in water and heated them for a form of hot pink lemonade. The Cahuillas used a tea of the leaves as a treatment for coughs and colds. Some reports say that young shoots were used raw and the dried leaves were smoked. White sap of the Sugar Bush which exudes from the

Fig. 11. LEMONADE BERRY *(Rhus integrifolia)*. Actual size.

fruit was used as an acid flavoring. The fruit is eaten by Mountain Quail, Crow, Red-shafted Flicker, Scrub Jay, and Cactus Wren. The twigs and foliage are browsed by Mountain Sheep. Today, foragers still enjoy the lemonade made from the soaked berries. The berries make pleasingly tart snacks if placed in the mouth right off the bush. The pulp is not swallowed but merely sucked for its juice.

22. Nuttall's Mariposa LILY
   (also Sego Lily) *Calochortus nuttallii* (pl. 2d)
   Family: Liliaceae

Description: An attractive member of the Lily family with a simple stem, 2-4 dm high. Linear leaves are reduced up the stem. Has 1-4 flowers, widely bell-shaped, white, tinged with lilac. The petals have a median green stripe and dark red or purple spots near the base. Petals 2.5-3.5 cm long. Blooms May through August.

Distribution and Habitat: Dry, bushy and grassy slopes and flats 5000-9000 ft (1500-2800 m) in Sagebrush Scrub, Pinyon-Juniper Woodland, east of Sierra Nevada, Lassen Co. to Inyo Co. A variety occurring in the Panamint Mountains does not have purple spots at the base of the petals. There are thirty-six species of *Calochortus* throughout the state. They range from white to brilliant orange and purple. The bulbs (technically corms) are very nutritious and tasty. They bear common names of Star Tulip, Globe Lily, Fairy Lantern.

Uses—Past and Present: Mormons in their first years in Utah consumed the bulbs in great quantities. This may be why it is Utah's state flower. Indians dug the bulbs and boiled, roasted, or steamed them in pits. Some tribes pounded the dried bulbs into flour and ate it as a mush. The bulbs preserve well. The entire plant has been used as a potherb. The seeds can be ground into meal, buds used raw in salads, and even the flowers eaten. These beautiful wildflowers should only be sampled if they are abundant. Several species are endangered and should not be used at all. Please assure positive identification and check the California Native Plant Society's rare and endangered plant list.

23. Bigberry MANZANITA,
   Arctostaphylos glauca (pl. 2e) Family: Ericaceae

Description: Manzanita is familiar to anyone who travels about the foothills or mountains because of its beautiful red exfoliating bark. It is a large evergreen shrub 2-4 m high, with gray-green leaves that are oval and 2-4.5 cm long. Urn-shaped flowers are white (or with some pink) and 8-9 mm long. Fruit is round, sticky, 12-15 mm in diameter, resembling a small apple from which its name comes. Fruit is reddish-brown when ripe. Blooms December through March; fruit appears in spring.

Distribution and Habitat: Common on dry slopes below 4500 ft (1400 m) in Chaparral and coastal mountain ranges of southern California. Inner coastal mountains to Mt. Diablo and Contra Costa Co. Though the Bigberry Manzanita is most abundant in the south, all of the species of *Arctostaphylos* can be used in the same way. The Greenleaf Manzanita, *A. patula,* is prized for its berries. It is found in pine forests from Kern Co. north in the Sierra Nevada, and the coastal mountains from Lake Co. north. The Parry Manzanita, *A. manzanita,* is also relished for its berries. It occurs in Chaparral, Foothill Woodland, Northern Oak Woodland, and Yellow Pine Forest of the interior coastal ranges from Contra Costa Co. to Humboldt, Trinity, Shasta, and Lake counties, and the foothills of the Sierra Nevada from Mariposa Co. north.

Uses—Past and Present: There are about forty-three species of manzanita throughout California, most of which were used for food by Indians and early settlers. The Spanish preferred the green berries to make a drink or jelly, while the Indians usually collected the ripe fruit. Several tribes celebrated the ripening of the manzanita with a big feast and dance. They collected the fruit by handpicking or shaking it into collecting baskets. The Miwoks ground up the berries and poured hot water over them in a strainer basket to extract the flavor which resembles cider. In some tribes the seeds were separated and ground into flour. The leaves were used by other tribes to make a wash or lotion for Poison Oak, a drink for headache, or a poultice for sores. Karoks and

Cahuillas used the wood for utensils and pipes. Among the Cahuillas, tracts of manzanita were owned by families. It served as an indicator of wild game, especially deer, mountain sheep, and coyotes. The green berries are made into jelly and beverages. The blossoms of some species are also steeped for tea, while the burls at the base of the shrubs are cherished for their beautiful wood products.

Squirrels, chipmunks, kangaroo rats, and numerous birds are drawn to the thickets to feed. The dense evergreen foliage provides shelter for the birds and smaller animals. Manzanita also acts as a valued erosion retardant and some species can be seen sprouting from the burls after a fire. Many species of *Arctostaphylos* are on the rare and endangered plant list of the California Native Plant Society and should be used only in emergencies.

MANZANITA CIDER

green manzanita berries   water
sugar or honey

Cover green berries with water in a saucepan and simmer 15 minutes or until somewhat soft. Bruise the berries but do not crush. Let stand overnight. Decant the liquid, let sediment settle and decant again. Sweeten if desired.

MANZANITA JELLY

1/2 gal manzanita   1 cinnamon stick
    berries, ripe or green   4 cups sugar
sliced peel of 1/2 lemon

Treat the berries as you would apples and follow any recipe for apple jelly or cover berries with water and crush. Add lemon peel and cinnamon stick, and simmer for 15 minutes. Place in cheesecloth or jelly bag and strain. Bring the juice back to a boil and for each 5 cups of juice, add 4 cups of sugar and boil rapidly until the liquid sheets, rather than drops, off the spoon. Pour into sterilized jars and seal. A drop of green food coloring brightens the jelly. Makes 5 cups.

24. Showy MILKWEED
Asclepias speciosa (fig. 12) Family: Asclepiadaceae

Description: Stout herb 5-12 dm tall, leafy to the top.
Leaves opposite, oval, 8-15 cm long. Flowers occur in umbels
with surrounding short hair. Fleshy petals are rose to purple
in color, becoming yellowish with age. Milky juice in all plant
parts. Fruit is densely wooly, 6-10 cm long in a pear shaped
pod, with short, soft spines. Blooms May through July, with
fruit following.

Distribution and Habitat: Dry gravelly and stony places
below 6000 ft (1800 m). Yellow Pine Forest and Mixed
Evergreen Forest of coastal ranges from Solano to Siskiyou
counties, west base of the Sierra Nevada south to Fresno Co.,
eastern slope from Inyo Co. to Nevada. Frequent in
disturbed areas.

Uses—Past and Present: The California and other Indians
used milkweed in many ways. After extracting the bitter and
possibly poisonous sap in several changes of boiling water,
they ate the stems, leaves, flowers, and young fruits. The
Cahuillas obtained a fine fiber from the stem which was used
for nets, slings, and snares. The seeds were ground into flour
or made into a remedy for rattlesnake bite or saddle sores on
horses. One account records that the Yuki Indians of
Mendocino Co. ate the blossoms raw. Other reports indicate
the roots were boiled and eaten with meat. The gum was
sometimes mixed with fat, hardened, and used for chewing
gum.

The most frequent use of milkweed by the various tribes,
however, seems to have been as a medicinal herb. They
applied the fresh sap daily for several days, to remove warts.
The Indians also applied this treatment to ringworm. The sap
was used as a healing agent for cuts and wounds and for
tattooing. Some tribes crushed milkweed leaves and made
them into a tea for measles, coughs, and swellings, while
Navajo women drank a tea prepared of the entire plant to
assist them in recovering from childbirth. To promote
temporary sterility, Indians of Canada drank an infusion of

Fig. 12. Showy MILKWEED *(Asclepias speciosa).* 1/2 X.

the pounded roots of *A. syriaca.* Other tribes drank this for rheumatism, pneumonia, or dysentery. Omaha Indians chewed the fresh root of one species, *A. tuberosa,* for bronchitis and other respiratory complaints. It was entered in the U.S. pharmacopoeia for treatment of various chest and lung ailments until 1936.

Monarch butterfly caterpillars use the genus as a food source. Unopened flowers and other young parts, such as leaves or fruit, though often reported poisonous, are fine vegetables when cooked. They should be placed in boiling water and cooked for one minute; this process repeated three times. Discard the water each time. Generally, the narrow-leaved milkweeds have been shown to be toxic, and caution should be used when approaching any new species. The broad-leaved species, *A. speciosa* and *A. syriaca,* have been eaten by many foragers with no ill effects. *A. syriaca* occurs primarily in deserts.

BUTTERED MILKWEED PODS

water	1/2 tsp salt
2 cups young	dash of pepper
milkweed pods	1/2 cup grated cheese
1/4 cup butter	

Choose young pods that are still hard to the touch when squeezed. Bring a pot of water to a boil, add the pods, and boil for 1 minute. Drain. Repeat this process three times, until pods are tender. (Cold water sets the bitterness, so water must be boiling prior to adding the milkweed.) Melt butter with salt and pepper. Add a little cheese if desired. Toss this with the cooked milkweed. Serves 4.

This recipe may be used for young stems, leaves, or buds, too.

25. MINER'S LETTUCE

(also Indian Lettuce) *Claytonia perfoliata* formerly *Montia perfoliata* (pl. 2f) Family: Portulacaceae

Description: A small, shiny, semisucculent annual herb, branched from the base and 1-3 dm high. Basal petioles or stems often tinged reddish. Lower leaves narrow or spatulate, but upper leaves form a disk which surrounds the stem out of which small white or pink flowers arise. Flowers have 5 petals 4-6 mm long. Blooms February through May. This has numerous variations in the basal leaves, but the upper disk and dainty flowers make it easy to recognize.

Distribution and Habitat: Common in shaded and moist places below 7500 ft (2300 m) throughout Coastal Sage Scrub, Chaparral, Foothill Woodland, and forests of most of the state as well as drier areas of the mountains of the Mojave Desert.

Uses—Past and Present: Indians ate the tender, fleshy leaves raw or cooked and made a tea of the plant to use as a laxative. Miners used the leaves—an important source of vitamin C—as salad greens to help prevent scurvy.

This plant is one of our few native plants introduced into

other countries. It has been taken to Europe where it is cultivated under the name Winter Purslane and is used for salads and as a potherb as it is here. It also grows wild in Cuba, where it was introduced. The black, shiny seeds are an important source of food for songbirds such as the Mourning Dove, Lazuli Bunting, Oregon Junco, and California Horned Lark. The kangaroo rat eats the leaves as well as the seeds.

A tasty salad can be made from Miner's Lettuce by treating the greens as you would lettuce, adding other vegetables and seasonings. It is especially nice with tangy ingredients, for it is rather bland by itself. Its slight succulence makes pleasant chewing. Backpackers like to mix Miner's Lettuce, Sheep Sorrel or Mountain Sorrel, and wild onions for a gourmet salad. It is also an excellent potherb when cooked like spinach.

26. Wild MINT
(also Field Mint) *Mentha arvensis* (fig. 13)
Family: Labiatae

Description: Familiar because it is frequently cultivated, an herb 1-8 dm high with opposite, light green leaves 2-7 cm long. Stems are square, as are stems of most mints. Flowers, lilac to purple, 5-6 mm long in the axils of leaves. Blooms July through October.

Distribution and Habitat: Moist places below 7500 ft (2300 m) in most plant communities throughout the state. Probably our only native wild mint.

Uses—Past and Present: In the eastern United States one tribe of Indians roasted the leaves over a fire and ate them with salt. The Cheyennes prepared a decoction of the ground leaves and stems and drank the liquid to check nausea. Southwestern Indian tribes flavored their cornmeal mush with mint, and some today still bake their fish in mint leaves. Besides its historical uses in jelly and as a flavoring, the fresh leaves make an agreeable mouthwash, leaving a fresh taste after an initial sensation of heat. Menthol is derived from cultivated *M. arvensis*. Tea from the fresh or dried leaves is enjoyed by campers and is a common remedy for nausea,

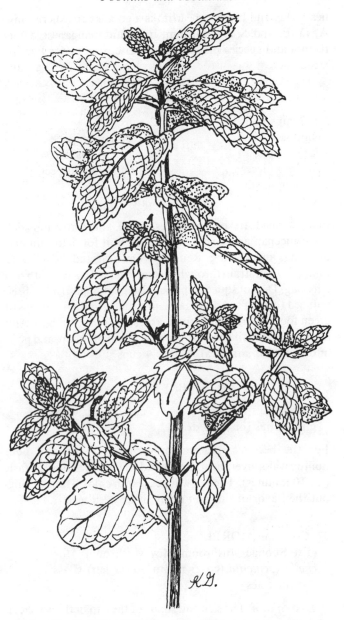

Fig. 13. Wild MINT *(Mentha arvensis).* Actual size.

headache, and heartburn. Mint is a good source of vitamins A, D, E, and K and calcium, iron, and manganese. Other recipes and species of *Mentha* are discussed under Peppermint.

MINT JELLY

2 1/2 cups mint leaves, chopped
1 large apple, cut into pieces
1 cinnamon stick (optional)

2 cups sugar
2 Tbsp lemon juice
green food coloring
1/2 bottle liquid pectin (6 oz. size)
water

Place washed, fresh mint leaves, apple, and cinnamon stick in a saucepan and cover with water. Boil for 5-10 minutes, mashing, and stirring. Remove from heat and let stand 15 minutes, then strain through several layers of cheesecloth or a jelly bag. Do not squeeze the bag. Boil liquid until it has been reduced to 2 cups. Add sugar, lemon juice, and a few drops of green food coloring, if desired, and bring to a hard boil. Add liquid pectin, boil hard for 1 minute, skim off foam, and pour into sterilized jars. Seal. Makes 4 cups.

MINT TEA

1 Tbsp fresh finely cut mint sprigs (or 1 tsp dried)

1 cup boiling water

For the best tea, place mint sprigs in a cup and pour boiling water over them. Cover and let it steep to desired taste (try 10 minutes). The tea will be clear in color. Honey brings out the flavor of this and many other herb teas.

27. Common MOREL
    (also Sponge Mushroom, May Mushroom) *Morchella esculenta* (cover photo, lower left) Class: Ascomycetes

Description: The size and shape of the cap in this species is variable, but the irregular pitted and ridged condition of the surface of the cap is very characteristic and reminiscent of a coarse sponge. The roughly conical cap is globe-to-top-shaped, 5-9 cm long and 3-6 cm thick. The pits are irregular in shape, 3-6 mm in diameter and up to 15 mm long, light tan in color. The spore-bearing cells are yellowish to yellow-brown or gray-brown. The stem that holds up the cap (actually a stipe) is short and stout, 1-3 dm long, whitish to slightly discolored, and somewhat wrinkled at the base. The cap and stipe are both hollow. A similar species commonly confused with this one is *M. conica,* which has darker pits and lighter ridges and is not quite as large. Another species, *M. angusticeps,* has blackish ribs on the cap. All of the morels are edible, but as in the use of all fungi, care must be taken to assure that it is correctly identified. Particularly avoid morel-like fungi in summer and fall which may be "false morels."

Distribution and Habitat: M. conica is commonly found after rains in January among ice plant in sandy soil of coastal regions. *M. esculenta* occurs in the spring (commonly in May) in wet abandoned meadows, orchards, and foothills where there is plenty of humus. *M. angusticeps* appears in the higher coniferous forests. Because of their characteristic pitted and ribbed cap, these are among the easiest of fungi to identify. Beware of the fungi with rounded or irregular caps that have wrinkled, folded, or smooth surfaces instead of the characteristic cross-linked ridges and pits of the morel.

Uses—Past and Present: To the ancient Greeks and Romans, the origin of fungi was a mystery. They marvelled at the existence of these plants without seeds or visible roots. A common belief was that they are produced by thunder which accompanies a rainstorm, a belief still held in parts of the Philippines. The American Indians seldom ate the fungi, perhaps because of the difficulty in differentiating the poisonous from the edible. Some morels were collected off the cottonwood trees in Cahuilla territory and prized for frying and making gravy or adding to acorn mush.

Morels should be cooked prior to eating—the fresh ones have made some people ill (though not seriously). When cooked, this is one of the most sought-after members of the edible fungi and is best cooked in butter or stuffed. Their hollow centers make them excellent candidates for stuffing with chicken or other meats and vegetables. Other recipes will be found under Puffball and Shaggy Mane.

STUFFED MORELS

2-3 doz fresh morels
1/2 cup butter
  or margarine
1/4 cup finely chopped
  green pepper

1/4 cup finely chopped
  onion
1 Tbsp chopped parsley
1/2 tsp minced garlic
1 egg, well beaten

Preheat oven to 350° (mod.) Rinse mushrooms and make a slit down the length of the cap. Chop stems and set aside. Melt 1/4 cup butter or margarine in a skillet and sauté caps 2 minutes. Arrange in shallow baking dish, slit side up. Melt remaining 1/4 cup butter or margarine in the skillet, add chopped stems, green pepper, onion, parsley, and garlic and cook until tender (about 5 minutes). Remove from heat and stir in egg. Heap mixture in caps and bake for 25 minutes. Serves 5. Optional additions are grated cheese and chopped meat.

CREAM OF MUSHROOM SOUP

2 cups milk
4 Tbs butter
  or margarine
3 Tbsp flour
2 cups sliced mushrooms
2 cups cream (or milk)

1/2 tsp salt
1/8 tsp white pepper
1 tsp chopped fresh
  parsley
1 Tbsp dry white wine
  or sherry

Heat 2 cups of milk to a simmer (do not boil). In a 2-quart saucepan melt 2 tablespoons of butter over low heat. Add 3 tablespoons of flour and blend. Cook this for 2 minutes, stirring constantly, and add about 1/2 cup of the hot milk. Blend and add remaining hot milk, stirring to keep the mixture smooth. Simmer for 3 minutes. In a skillet, heat 2 tablespoons of butter until bubbly and cook the sliced mushrooms in it for 3 minutes, stirring often. Add the mushrooms to the simmering sauce. Add remaining cream or milk and seasonings and simmer for about 3 minutes. Serves 4-6.

28. Tansy MUSTARD
(also Peppergrass) *Descurainea pinnata* (fig. 14)
Family: Cruciferae

Description: Pubescent annual 1-6 dm tall. Deeply incised leaves 3-9 cm long. Petals yellow, 4 in number, with 6 stamens. Petals 1.5-2.5 mm long. Fruit a slender pod 5-12 mm long, usually curved. Seeds reddish-brown, round, and flat. Blooms March through June.

Distribution and Habitat: Dry sandy and disturbed rocky places below 8000 ft (2500 m). Many plant communities along the coast from Contra Costa Co. to San Diego, San Joaquin Valley, southern Sierra Nevada, Mojave and Colorado Deserts. A genus of five species and numerous varieties, all of which are edible.

Uses—Past and Present: Seeds of many of the wild mustards were used by Indians. They were gathered by knocking them into baskets, then parched and ground for use in soup or mush. They were also used crushed as poultices or made into tea for "summer complaint." *D. pinnata* is sold in Mexican drugstores as *Pamito.* The Pomo Indians mixed the seeds with their cornmeal, while the Cahuillas ground the seeds for stomach ailments.

The leaves can be boiled or roasted and eaten, though several changes of water are necessary to make them palatable. It has been noted that large quantities of fresh leaves fed to cattle over long periods of time have caused poisoning. This is also true, however, of many of our cultivated mustards, and also cabbage, broccoli, and cauliflower. Therefore, moderate quantities are considered safe for humans and to my knowledge no human poisoning has ever been recorded for this species.

Fig. 14. Tansy MUSTARD *(Descurainea pinnata).* 1/2 X.

29. Coast Live OAK
(also Encina) *Quercus agrifolia* (fig. 15) Family:
Fagaceae

Description: Broad evergreen tree. Leaves convex on
upper surface, oval to elliptic, usually sharp or spiny-toothed,
mostly 3.5-7 cm long, similar in appearance to holly leaves.
Fruit an acorn, 2.5-3.5 cm long. Leaves remain on branches
for one season and are discarded when new foliage is
produced in spring.

Distribution and Habitat: Mostly restricted to canyon
bottoms and humid hillsides of outer coast ranges to middle
of the state and western slopes of inner ranges. Occurs below

3000 ft (900 m) in Southern Oak Woodland, Oak and Foothill Woodland and occasionally in cultivated and urban areas. There are sixteen species of oak recognized in California, with numerous hybrids. Some of the more common and desirable for foragers are: *Q. douglasii,* the Blue Oak; *Q. chrysolepis,* the Canyon Oak; *Q. kelloggii,* the California Black Oak; and *Q. lobata,* the Valley Oak or Roble.

Uses—Past and Present: Before acorns can be used, the bitter and constipating tannin must be leached out. This can be accomplished by rinsing chopped or ground nuts in water until they are no longer astringent. (See Introduction.) Some of the nuts are naturally sweet and do not require this process. California Black Oak, *Q. kelloggii,* is one of the sweeter acorns, but does require some leaching. I have found the cultivated Cork Oak, *Q. suber,* and the Holly-leaved Oak, *Q. ilex,* to be excellent when roasted.

Acorns were of primary importance to many Indians. They often placed the ground nuts in sand depressions alongside riverbeds or used a basket and allowed the water to percolate through the mass until the astringent taste had disappeared. The central meal was taken from the sandy depression and reserved for bread because it was free of sand. The remainder of the dough was mixed with water to form a soup. The Concows of Round Valley mixed the bread dough with red clay before baking, claiming this made it sweeter. The acorn bread became black when baked and soon dried to a very hard loaf. John Muir was fond of this bread because it was such a compact, strength-giving food.

Pomo Indians considered the oaks to be personal property and passed down the possession of the trees in the family with definite rules. Some tribes also discovered the precursor to today's penicillin-type drugs: the ground acorn meal was allowed to accumulate mold, which was scraped off to use for boils, sores, and inflammations. The wood ashes were used medicinally among the Cahuillas. Dye was made from the bark and tannin was used for curing buckskin. Whole acorns were also used to make musical instruments and necklaces as well as toys and trade items.

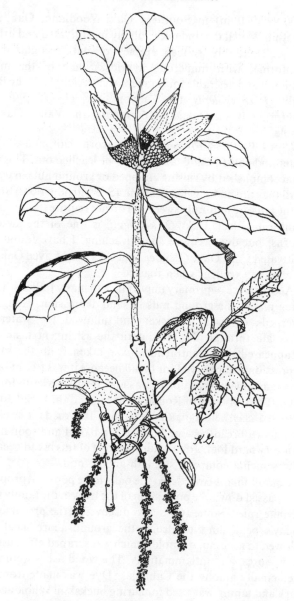

Fig. 15. Coast Live OAK *(Quercus agrifolia).* Actual size.

Today, the oak is still prized for its wood and cooling shade. As they were a staple food for the Indians, acorns are still of paramount importance to many species of wildlife. The acorn's greatest value is its availability during the winter when other food is scarce. Quail devour small acorns as do squirrels, chipmunks, deer, elk, mountain sheep, and Black Bear.

People who use acorns today agree that they resemble other nuts in oiliness and flavor. They contain significant quantities of calcium, magnesium, phosphorus, potassium, sulfur, fat, and protein. They are especially good in cookies, breads, and pies. This is one of my favorite recipes.

ACORN ROCA BARS

1 cup butter
  or margarine
1 cup brown sugar
1 egg
1 tsp vanilla
2 cups flour

1/2 tsp salt
3/4 cup finely chopped,
  leached acorns
12 oz milk chocolate or semi-
  sweet chocolate pieces
1/2 cup sweetened
  shredded coconut

Preheat oven to 350° (mod.). Cream together butter or margarine and brown sugar. Blend in egg and vanilla. Add flour and salt. Stir in 1/2 cup acorns and spread the thick mixture in an ungreased 10x15-in pan. Bake for 40 minutes. Remove from oven and spread milk chocolate or semi-sweet chocolate pieces over the cookie mixture, smoothing as it melts. Mix together 1/4 cup acorns and coconut and sprinkle this on top of the melted chocolate. Cut and cool.

Recipes for Walnuts and Hazelnuts may also employ acorns.

30. Wild Red ONION
  (also Red-Skinned Onion) *Allium haematochiton*
  (fig. 16) Family: Liliaceae

Description: All onions are herbs arising from bulbs. Wild Red Onion bulbs, 2-3 cm long, have a deep red or white coating. Leaves are mostly basal, with the unmistakable taste and smell of onion. Flowers small, in clusters of 10-30 per terminal umbel, white to rose with darker midvein. Blooms March through May.

Distribution and Habitat: Along watercourses or dry slopes and ridges of clay or stony soil below 2500 ft (800 m), Chaparral, Valley Grassland, and Coastal Sage Scrub from San Luis Obispo Co. to Baja California. All thirty-eight species of *Allium* in California, in various habitats, are easily recognized by their general resemblance to cultivated onions in both appearance and smell. The genus includes garlic, leek, and chives. In wet mountain meadows, the Swamp Onion, *A. validum,* is particularly abundant. There are many "Onion Valleys" in the Sierra due to this and other species.

Uses—Past and Present: Slaves who built the pyramids of Egypt were fed onions to prevent scurvy. In America the town of Chicago derives its name from the Indian name for leeks, which grew in the original location. Besides using the onion for food, Indians obtained a red dye from its skin and used the crushed bulbs of garlics and onions to relieve the pain of insect stings and bites. Bulbs were usually eaten raw but were often roasted to improve the flavor. The juice can be boiled down to a thick syrup and has been used as a treatment for colds and throat irritations. Whether using for flavoring or food, make sure the plant smells like onion—the Death-Camas (which does not have the smell and has white flowers that are not in an umbel) grows in the same areas as some of the mountain onions (see pl. 8f). The Golden-Mantled Ground Squirrel includes the onion in its diet. Several species of *Allium* throughout the state are on the rare and endangered plant list and should be used only in extreme emergencies.

CREAMED WILD ONIONS

1 cup onion bulbs and leaves
1 Tbsp butter or margarine
2 Tbsp flour
3/4 tsp salt
dash of pepper
1 cup milk

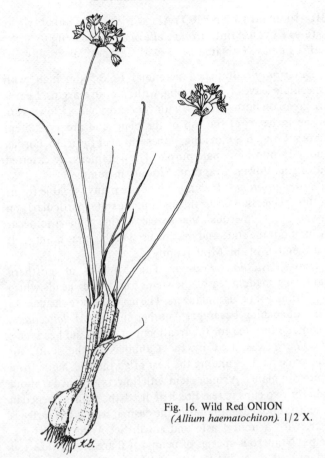

Fig. 16. Wild Red ONION
(Allium haematochiton). 1/2 X.

Remove the outer skins and tough leaves from the plants. Tear the tender leaves into bite-sized pieces and simmer all the onion in a little water until tender. Melt butter or margarine. Add flour, salt, and pepper and cook, stirring constantly, for 1 minute. Add cooked onion. Stir in milk and simmer until thickened.

This basic recipe can be made with backpacker's provisions such as powdered milk and any vegetable at hand. Fish may be added along with more milk to make soup or chowder.

31. Mountain PENNYROYAL
(also Coyote Mint) *Monardella odoratissima* (fig. 17) Family: Labiatae

Description: Branched perennial 1.5-3.5 dm high, with lanceolate leaves 1.5-3 cm long, with smooth margins. Leaves are green on both surfaces, with very short hairs. Flowers are borne in heads at the top of the plant and are 1.5-2.5 cm across. Calyx 6-8 mm long, pubescent near the teeth, corolla about 15 mm long, pale purple. The 4 stamens are exserted out of the flower. Fragrant. Flowers in summer.

Distribution and Habitat: There are many variable forms of this species that occur throughout the state, particularly on dry slopes. Numerous other species of this genus also occur throughout the state and may be used in the same manner. It is a member of the Mint Family.

Uses—Past and Present: The Miwoks of northern California made a tea of the stems and flower heads which they drank for colds and fevers. The tea was also considered a thirst-quenching beverage. Another species, *M. lanceolata,* was made into tea for the treatment of colds and headache; *M. villosa* was used by the Cahuillas in the south for stomachache. It is among the best of all the teas made from mountain plants. A single stem with flowers attached is about right for two cups of tea. Fresh or dried, the stem is steeped in water (not boiled) until it is of the desired strength. The tea is clear, with a minty taste and smell. See Wild Mint for a recipe. Numerous species of pennyroyal are on the rare and endangered plant list and should be used only in extreme emergencies.

32. PEPPERMINT
Mentha piperita (fig. 18)
SPEARMINT
Mentha spicata Family: Labiatae

Description: Fragrant herbs with opposite leaves and square stems. Stems often turning purplish. Flowers are rose-purple to white and occur July through October. Peppermint

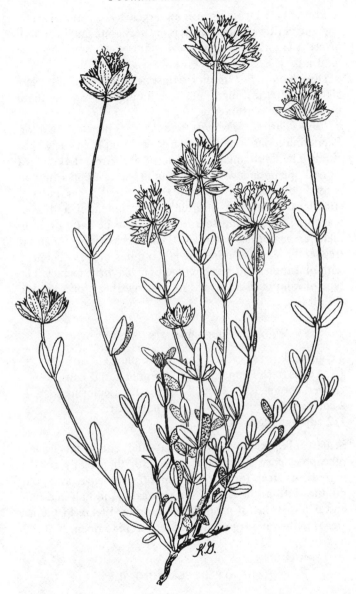

Fig. 17. Mountain PENNYROYAL *(Monardella odoratissima)*. Actual size.

is differentiated from Spearmint largely by its petioled leaves and longer flowering spikes. It is not very abundant in natural places. These are introduced mints but have become naturalized.

Distribution and Habitat: Moist fields and marshes below 5000 ft (1500 m), many plant communities from southern California to Washington.

Uses—Past and Present: Because of its penetrating odor, Peppermint was used to detect leaks in pipes in early U.S. cities. The Menominee Indians of the Great Lakes area treated pneumonia with a tea of mint leaves. Peppermint oil is used commercially to relieve excess stomach gas and as a stimulant. It is also used in candy and gum and as a seasoning. Spearmint does not appear to be naturalized in many areas but has commercial uses in chewing gum and is frequently used for mint jelly. The mints blend deliciously with pineapple, and here are two of my favorite recipes. The type of mint used in each is interchangeable. For a tea, see Wild Mint.

PINEAPPLE-PEPPERMINT SHERBET

2 cups milk
1 1/4 cups fresh Peppermint
  leaves, chopped
2 cups crushed pineapple
1/2 cup light corn syrup

1 cup plus 2 Tbsp sugar
2 cups half-and-half
green food coloring
  (optional)

Scald milk. Remove from heat and add mint leaves, pineapple, corn syrup, and sugar. Stir thoroughly and let ingredients steep for 1 hour. Skim off mint leaves and pour mixture into an ice cream freezing can. Add half-and-half and a few drops of green food coloring. Freeze in the ice cream maker and transfer to a deep freeze to ripen. Makes 2 quarts.

PINEAPPLE-SPEARMINT ADE

1 1/4 cups Spearmint
  sprigs, tightly packed

1 cup boiling water
6 Tbsp honey or sugar

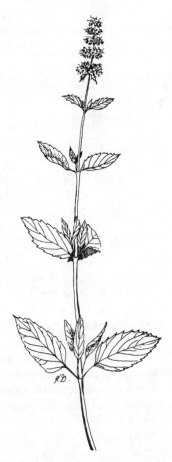

Fig. 18. PEPPERMINT *(Mentha piperita)*. 1/2 X.

1/4 cup lemon juice
1 cup pineapple juice,
  chilled

1 (28 oz.) bottle club soda
Green food coloring
(optional)

Wash Spearmint and place in saucepan with water and honey. Simmer, uncovered, 10 min. Chill. Strain and add lemon juice, pineapple juice and club soda to the strained liquid. Tint as desired with green food color. Serve over crushed ice. Makes 6 to 8 servings.

33. Sugar PINE
Pinus lambertiana (fig. 19) Family: Pinaceae

Description: The largest of our pines, with needles 7-10 cm long, in bunches of five. Cones 2.5-4.5 dm long, hanging on ends of branches alone or in twos or threes. Pine nuts and seeds are usually produced in the fall. The cultivated Aleppo Pine, *P. halepensis,* has a thin-shelled nut of excellent quality. The shell can be eaten.

Distribution and Habitat: Common forest tree from 2500-9000 ft (800-2800 m), southern California to Oregon mountains. One of a large genus (twenty-one species) which has edible nuts. Those species with large sweet nuts are *P. coulteri,* Coulter Pine, producing the heaviest cones of all the pines; *P. sabiniana,* Digger Pine, which has a forked trunk and a cone similar to Coulter; and *P. albicaulis,* Whitebark Pine, found at higher altitudes in the Sierra Nevada.

Uses—Past and Present: Practically all parts of the pine trees of the state were used in some way by various Indian tribes. Seeds were usually roasted. Karok Indians used the seeds of Digger Pine strung on thread made of Wild Iris fibers to decorate dresses worn in their dances. Charcoal left from burning nutmeats was used by other tribes to treat sores and burns. The soft center of green cones was roasted in hot ashes, then eaten. The pitch was a valuable glue for mending canoes and fastening arrowheads and feathers. It was also smeared on burns and cuts or chewed like gum to allay rheumatic pains. Twigs and rootlets yielded material for sewing baskets. Medicinal teas were made from twigs, leaves, and sometimes the bark. The inner layer of bark was so frequently employed as an emergency food that great stands of these trees were found stripped of their bark by early settlers. The Mohawk name *Adirhondak* describes a group of eastern Indians who were "tree-eaters." These people ate quantities of the inner bark from the top of the pine. The Zuni prepared the bark by scraping it with a sharpened stone or animal horn, boiling it, pounding it into a mash, and shaping it into cakes. These were cooked in a stone-lined baking pit, then smoked. Prior to eating it they softened the cakes by boiling them in water.

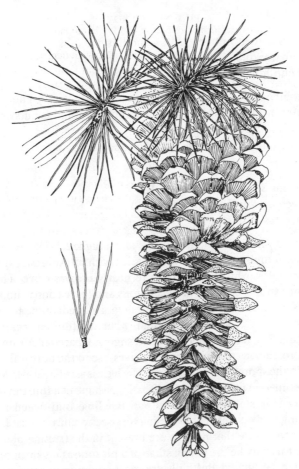

Fig. 19. Sugar PINE *(Pinus lambertiana).* 1/4 X.

Foragers today still enjoy tea from pine needles, and the fresh or roasted nuts are a delicacy. Pines rank near the top of the list in importance to wildlife. Many birds and mammals feed on the nutritious, oily seeds. The needles are eaten by the Mule Deer and other browsers, which include Porcupines and small rodents. The trees provide good cover all year round. Pines provide nesting places for the Mourning Dove and many other birds.

PINE NUT CANDY

3 Tbsp melted butter
  or margarine
1/2 cup honey

nonfat dry milk
3/4 cup finely chopped
  pine nuts

Combine butter or margarine and honey. Stir in enough nonfat dry milk to make a thick mixture. Add pine nuts and form into a roll. Chill and slice.

Pine nuts may be substituted in recipes for Pinyon.

34. Gem-Studded PUFFBALL
Lycoperdon perlatum (fig. 20) Class: Basidio-mycetes

Description: Fruiting body, 3-7.5 cm high, 2.5-7 cm wide, ovoid to top-shaped, the outer coating with soft spines, scales or granules, whitish or grayish to nearly brown in age. The stouter spines or warts are deciduous and leave conspicuous spots when they fall off. The spore mass in the fruiting body turns yellow, then olive-colored when mature or nearly mature. The fruiting body is often slightly depressed around the protrusion, which eventually gives rise to the pore out of which the spores will be released. The base is small and sterile, sometimes giving the entire puffball the shape of a top, but in others it is stemlike. The puffball is edible only when its fruiting body is entirely white in cross-section. Care must be taken to ensure that no gills are present in the fruiting body since this may be the young stage of a poisonous mushroom. The Gem-Studded Puffballs are most common in spring.

Distribution and Habitat: Throughout the state. Single or in clusters on decayed wood or humus of forested regions, but also common in urban areas where the soil is moist and heavily organic. They are also known to occur in lawns and pastures as well as meadows.

Uses—Past and Present: The Miwoks ate puffballs, as did the Zunis and Iroquois. At pueblos along the Rio Grande Valley, Indians sometimes treat broken eardrums with powdered puffball. Spores of the mature fungus are sprinkled over wounds to help stop the flow of blood.

Fig. 20. Gem-Studded PUFFBALL *(Lycoperdon perlatum).* Actual size.

The white-meated puffballs (all of which are edible) are very tasty in casseroles or added to vegetable or meat dishes. The rule is that any puffball that is white and uniformly smooth in texture throughout its inside is good to eat. Slice the puffball in half to be sure there are no gills or colors other than white. Each specimen should be checked, since edible puffballs may be found near the button stages of poisonous mushrooms. The edible puffballs also have a mild "earthy," mushroomlike odor.

Puffballs are best preserved by drying in thin slices or sautéeing and freezing. Mushroom recipes abound, and puffballs can be used in most. Be sure to sauté puffballs lightly in butter and do not overcook them. Other recipes will be found under Morel and Shaggy Mane.

SOUR CREAM PUFFBALLS ON TOAST

1 medium onion, chopped	1/2 tsp salt
2 cups sliced puffballs	1/4 tsp pepper
1 Tbsp butter or margarine	1 cup sour cream
	6 slices toast

Sauté onion in butter until clear and tender, adding more butter if necessary. Add puffballs, salt, and pepper and sauté until tender. Stir in sour cream with heat on lowest setting possible, and serve over toast at once. Serves 6.

35. White-Stemmed RASPBERRY
Rubus leucodermis (pl. 3a) Family: Rosaceae

Description: Thorny, trailing shrub with stems about 2 m long, rooting at tips and forming thickets. Stems have heavy whitish bloom when young. Leaves 3-5 foliate, green above, whitish below. Leaflets oval, the terminal 7-9 cm long. White flowers mostly 3-10 per cluster on lateral shoots, 7-10 mm across. Fruit dark purple or blackish, sometimes yellow-red, coming away from stem in a cone with a hollow center. Fruit 1.5 cm in diameter. Blooms April through July with fruit following.

Distribution and Habitat: Open forest slopes and canyons, often near streams below 7000 ft (2100 m). In Sierra Nevada from Tulare Co. north and in coastal mountain ranges from Santa Cruz Co. north to Siskiyou and Del Norte counties. A variety which occurs on dry flats and slopes is found in the San Bernardino and San Gabriel mountains and San Diego Co.

Uses—Past and Present: A familiar berry, the raspberry is used in much the same way as the blackberry. Fruit is tangy and is much prized for jams and sauces. Refreshing tea can be made from the leaves, which are high in vitamin C. Tender young peeled sprouts and twigs of raspberries and blackberries are pleasant to chew. Valuable as a deer browse. For other animal uses, refer to Blackberry.

RASPBERRY HARD SAUCE

1/4 cup butter	dash vanilla flavoring
or margarine	1 1/4 cups raspberries
1 cup powdered sugar	slightly crushed
2 Tbsp cream or milk	

Cream butter. Add sugar slowly, beating constantly. Slowly add cream, vanilla and raspberries, beating constantly. Serve over shortcake or slightly sweetened rice for a real treat. Serves 4.

Raspberry jam can be made with any commercial recipe, using 4 cups of berries to 2 cups of sugar.

36. Wild ROSE
Rosa californica (pl. 3b) Family: Rosaceae

Description: Sprawling shrub with curved prickles. Leaves are alternate, compound into 5-7 leaflets. Leaflets oval 1-3.5 cm long with fine teeth along edges. Flowers resemble cultivated Rose, but are smaller, with only 5 petals, white to pink. Fruit oval, 8-16 mm long, 1-15 mm thick, developing below the flower, red when mature. Blooms May thru August with fruit following.

Distribution and Habitat: Fairly moist places, such as along streams and river banks below 6000 ft (1800 m) in many plant communities west of Sierra. There are nine species of *Rosa* in California, making it an easy plant to find. All have edible, though not always palatable, fruits and flowers. *R. woodsii,* with straight prickles, has an unusually pleasant fragrance.

Uses—Past and Present: The fruit of the Wild Rose, known as the hip, is said to contain more vitamin C, calcium, phosphorus, and iron than oranges. During World War II in England, the hips were gathered for their abundance of these essential vitamins and minerals. Hips can be dried for tea or used for jelly or sauce. The dry inner seeds must be removed. They are popular with backpackers because even the dried fruits left over from the previous season remain on the bush and can be soaked in water and used. They resemble a small, dry apple in appearance and taste. The flowers have long been used in folk recipes for butter, perfume, candy, jelly, and tea. The Cahuillas picked the buds and ate them raw or soaked the blossoms in water to make a drink. Some tribes made a tea from the roots for colds and from the leaves and fruits for pains and colic. The wood was used for arrow shafts and the fiber from the bark was used in making twine and other goods. Rose leaf tea is very pleasant when enjoyed around a campfire.

The hips remain on the shrubs throughout the winter and into spring, thus providing good forage in times of scarcity. The rose thickets provide good shelter and nesting sites. The Audubon Cottontail and the wood rat are among those western animals utilizing the rose.

BASIC ROSE HIP SYRUP

2 cups rose hips	2 Tbsp lemon juice
water	1 Tbsp cornstarch
1 cup sugar	(optional)

This is an easy recipe because you don't need to remove the seeds from the hip. Cut off the sepal end of the hips with scissors, cover hips with water, and boil until mushy. Strain off the juice, cover hips with water again and make a second extraction. For every 2 cups of resulting juice, add 1 cup of sugar and 2 Tbsp lemon juice and boil until it thickens, or add 1 Tbsp cornstarch dissolved in a little water, and cook to desired consistency. Pour into sterilized jars and seal. Excellent over ice cream or pancakes.

37. SALAL
Gaultheria shallon (pl. 3c) Family: Ericaceae

Description: Attractive, spreading evergreen shrub with dark green leathery leaves 3-10 cm long, finely toothed. Urn-shaped white or pink flowers are 8-10 mm long. Fruit dark purple, 7-8 mm broad, with a distinctive star-shaped depression on the end away from the stem. Several seeds. Blooms April through July, with fruit occurring in summer.

Distribution and Habitat: In forests and brushy areas below 2500 ft (800 m). Abundant near the coast from Santa Barbara to Del Norte counties. Uncommon in southern California. A smaller-leaved species with a prostrate habit, *G. ovatifolia,* has red fruit and is found in wet places at higher altitudes.

Uses—Past and Present: Indians of the Northwest mashed Salal berries and dried them for winter use. Today they are used to make excellent jam, jelly, and berry pie. The fruit, buds, and leaves are eaten by Blue Grouse and Bandtailed Pigeon. The fruit alone is eaten by the Wren-tit, Douglas Chickaree, and the Roosevelt Elk.

ONE-CRUST SALAL PIE

3 cups fully ripe	1 cup sugar
Salal berries	1/2 tsp ground cloves

1/2 cup chopped	3/4 cup flour
pecans or walnuts	1/2 cup plus 2 Tbsp
2 eggs	melted butter or
3/4 cup sugar	margarine

Preheat oven to 325° (slow mod.). Grease a 10-in pie plate. Pour Salal berries into it, sprinkle with sugar, cloves, and chopped pecans or walnuts, and stir. Beat eggs until lemon colored and light, and while still beating add, a little at a time, sugar, flour, and melted butter or margarine. When thoroughly mixed and smooth, spread over the berries. Bake for 45 minutes or until the crust is brown.

38. SALMONBERRY,
Rubus spectabilis (pl. 3d) Family: Rosaceae

Description: Branching, leafy shrub, 2-4 m high, with only a few prickles. Older stems have yellowish shredding bark. Leaves tri-foliate, deciduous, terminal leaflet 4-10 cm long. Showy flowers usually 1-3 per cluster. Petals red-purple, 1.5-2 cm long. Fruit round or conic, red or yellow-orange, 1.5-2 cm long, pulling away like a raspberry. Fruits in summer.

Distribution and Habitat: Moist places in wooded areas below 1000 ft (300 m). Santa Cruz Mountains to Del Norte Co.

Uses—Past and Present: Salmonberry is a delicious berry used much the same as blackberries and blueberries. It was usually used fresh by the Indians but dried for future meals, and often pounded with meat and fat to form pemmican. Northwestern Indians often ate the berries with half-dried salmon eggs, thus the common name. The young shoots make a pleasant potherb. This is a mild-tasting fruit and may be prepared in the same manner as blackberries and blueberries by merely adjusting the amount of sugar and spices to taste. For animal uses, refer to blackberry.

39. Western SERVICE BERRY
(also June Berry, Shad Berry) *Amelanchier pallida* (pl. 3e) Family: Rosaceae

Description: Deciduous shrub, 1-3 m tall. Simple, alternate leaves, lower side paler than upper, oval in shape usually 2-4 cm long and 1.5-2.5 cm wide. Leaves are typically toothed on the edges above the middle, while the lower half lacks teeth. Flowers are white, petals 8-11 mm long in clusters of 4-6, with about 15 stamens. Fruit roundish, applelike, purple-black, 4-6 mm in diameter with upper part of flower tube still attached. Berries ripen in summer and tend to dry on the branches. Blooms April through June and fruits in summer.

Distribution and Habitat: Dry, rocky slopes and flats below 11,000 ft (3400 m) in many plant communities, but especially in Montane Forests and along streams. Coast ranges and Sierra Nevada south to Kern and Ventura counties. In San Gabriel, Cuyamaca, and Palomar mountains. There are four other species of *Amelanchier* in California, all of which produce edible berries.

Uses—Past and Present: Indians dried large quantities of this fruit for winter and mixed it with pounded dried meat for pemmican. An eyewash was made from boiled, green inner bark. Midwestern Ute Indians preferred to use the fruit of some species before it turned red or purple. Service berries are good stewed or dried and make excellent pies. Dried berries can be substituted in recipes for currants or raisins. The wood makes good arrow shafts. Pheasant and other birds like the fruit and buds, while Mule Deer, elk, and other browsers feed on twigs and foliage.

SERVICE BERRY JAM

6 1/2 cups fresh service berries
1/2 cup lemon or orange juice

1 pkg powdered pectin
5 cups sugar

Crush fresh berries to yield 3 3/4 cups pulp. Add lemon or orange juice and powdered pectin. Place over high heat and stir until it boils hard. Stir in sugar and bring to a full boil for 1 minute. Remove from heat, skim off foam and stir for 15 minutes to prevent floating fruit. Pour into sterilized jars and seal. Substituting peaches for half the amount of berries makes a delicious jam also.

The dried berries are quite different from the fresh or stewed berries and may be used in muffins and pancakes like currants or blueberries. Service berries may be dried in a slow oven or in the sun in layers until they are raisinlike.

40. Mountain SORREL
Oxyria digyna (fig. 21) Family: Polygonaceae

Description: Low herbaceous perennial with sour juice. Leaves alternate but mostly basal. Flowers are small and round with long petioles, reddish or green with reddish tinges, 4-6 mm long in terminal clusters. Blooms July through September.

Distribution and Habitat: Rocky places 8000-13,000 ft (2500-4000 m), mostly above tree line in Sierra Nevada, White and Yolla Bolly mountains north.

Uses—Past and Present: This backpacker's salad plant is often found around old mining camps, suggesting it was frequently used by the gold seekers. Indians chopped up sorrel with other leaves such as Watercress and allowed the mixture to ferment to form a type of sauerkraut. It is a refreshing plant to use in sandwiches and salads or as a potherb. The sour taste is probably due to oxalates, which will sometimes cause poisoning to livestock. No incidents have been reported of human poisoning, and the plant is frequently used in moderation by mountain climbers and backpackers.

Chopped sorrel is delicious with cottage cheese or cooked with milk, onion, butter, and flour into a soup. Use your favorite recipe for cream of onion soup and add equal amounts of onion and sorrel leaves. Additional uses will be found under Sheep Sorrel.

41. SQUAW BUSH
(also Squawberry) *Rhus trilobata* (fig. 22) Family: Anacardiaceae

Description: Diffusely branched shrub 8-14 dm tall, strongly scented when crushed. Leaves 3-5 lobed, pubescent, terminal lobes ovate, 1-3 cm wide. Flowers yellowish in

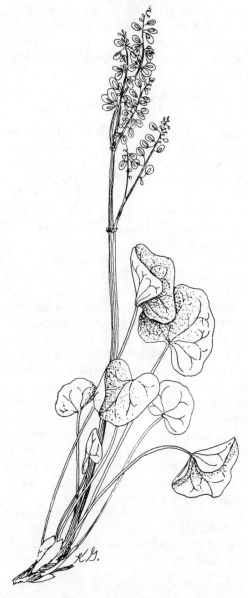

Fig. 21. Mountain SORREL *(Oxyria digyna)*. 3/4 X.

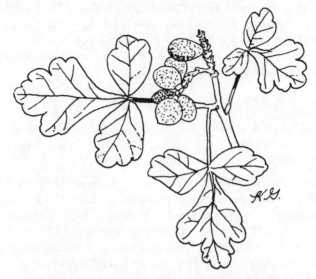

Fig. 22. SQUAW BUSH *(Rhus trilobata).* Actual size.

clustered spikes that appear before the leaves. Petals about 2 mm long. Fruit sticky, slightly pubescent, 4-5 mm in diameter, reddish with one seed. Leaves resemble those of Poison Oak, but Poison Oak has white fruit that appears after the leaves. Blooms March through April with fruit following.

Distribution and Habitat: Common in canyons, washes, dry slopes below 5500 ft (1700 m). In Coastal Sage Scrub, Foothill Woodland and Chaparral throughout southern California, coastal mountain ranges from San Diego to Siskiyou Co., Mojave Mountains, and northern Colorado Desert.

Uses—Past and Present: A member of the Sumac Family, the berries, were eaten fresh by Indians, ground into flour for soup, or soaked in water for a beverage. They were used medicinally for stomach ailments and powdered as a lotion for treating smallpox and sores. Hopis used the berries in making body paint for ceremonies and the wood as a fuel in their kivas. Navajos made a black dye from the roots. A tea from the stems was used to treat coughs. The strong, thin,

pliable stems were used in basketry by the Cahuillas. Luiseño
Indian women made a woven fan from the stems that was
used for beating seeds from grasses and other plants. For
more information regarding the use of Squaw Bush by man
and animals, see Lemonade Berry.

SQUAW BUSH SAUCE

1 1/3 cups water   1/2 cup dried, ground,
3/4 cup honey or   and sifted Squaw Bush
  1 cup sugar   berries

Use only the fine particles of berry that come through the
sifter; coarser particles may be soaked for juice. Combine all
ingredients, cover, and boil for 30 seconds. Reduce heat and
simmer for 15 minutes. Uncover and boil until mixture thickens, stirring constantly (about 10 minutes). It will thicken further when refrigerated. Pour into jar and refrigerate. Makes 1
cup. (Great on homemade bread!)

Fresh berries may be used if they are crushed, simmered,
and strained before addition of the sweetening.

42. Gairdner's SQUAW ROOT
    (also Yampa) *Perideridia gairdneri* (fig. 23)
    Family: Umbelliferae

Description: Erect branching herb from tuberous or
swollen roots which occur singly or in clusters. Plant 3-12 dm
tall, with leaves pinnate into linear segments 2-13 cm long.
Flowers occur in loose compound umbels; are tiny and
whitish. Blooms June through July.

Distribution and Habitat: Wet places below 11,000 ft (3400
m), many plant communities on west side of Sierra and in
mountains from San Diego Co. north. There are eight species
of squaw root in California, common in the Sierra. They are
all edible and occur in most habitats that offer moisture and
shade. One species, *P. pringlei,* occurs in drier Chaparral
from northern Los Angeles Co. to San Luis Obispo and Kern
counties. Care should be taken to get correct identification of

Fig. 23. Gairdner's SQUAW ROOT *(Perideridia gairdneri)*. 1/2 X.

this plant, since some of the members of the Carrot Family to which it belongs are poisonous (see pl. 8e).

Uses—Past and Present: Yampa tubers were such an important Indian food that towns, rivers, and valleys have been named after them. It was a close contender for the name of the state of Colorado, and it is believed that the Shoshone Indians were forced to attack stage coaches and the pony express when their grass and root foods were ravaged by the cattle and sheep brought in by the prospectors after the discovery of the Comstock Lode. Indians prepared the roots by placing them in water and tramping with their bare feet to remove the skin of the tubers, then washing and cooking them as a vegetable. Sometimes the roots were dried and then ground into flour for baking. One early report says that the seeds are good for colds and indigestion, while a poultice made of powdered seeds is good for the eyes and bruises. The fried herb and seeds were put hot into a bag to treat stomachaches. Blackfoot Indians chewed the root and swallowed the juice to relieve sore throat. The flavor of Yampa ranges from radishlike to carrotlike. The tubers can be used as you would parsnips or potatoes. Numerous species of *Perideridia* are on the rare and endangered plant list, including *P. gairdneri* and *P. parishii* from San Bernardino Co.

CREAMED YAMPA ROOT

1 cup yampa root
  tubers, chopped
1 Tbsp butter
  or margarine
2 Tbsp flour
3/4 tsp salt

dash of pepper
1 cup milk
1/4 cup freshly chopped
  mint, onion, or
  sorrel (optional)

Boil and peel tubers. Melt butter or margarine in a saucepan. Add flour, salt, and pepper and cook, stirring constantly, for 1 minute. Add tubers, freshly chopped mint, onion, or sorrel, and milk and simmer until thickened. Serves 2.

Yampa roots can also be candied like sweet potatoes by melting butter and brown sugar together and adding the cooked, peeled roots.

43. Wood STRAWBERRY

Fragaria vesca ssp. *californica* formerly *Fragaria californica* (cover photo, lower right) Family: Rosaceae

Description: Low plants with hardly any true stems showing above the ground. Familiar because of their resemblance to cultivated strawberries. Runners root at nodes. Leaves divided into 3 leaflets, terminal leaflet 2-5 cm long. Flowers white, petals 5-8 mm long, fruit a small strawberry, 1-1.5 cm thick. Blooms March through June. Bears fruit in summer.

Distribution and Habitat: Shaded, fairly damp places below 7000 ft (2100 m) in closed-cone forests and Chaparral from coastal mountain ranges, Santa Barbara to Del Norte counties and Siskiyou to Tulare counties in Sierra Nevada, as well as most mountains of southern California. There are four species of *Fragaria* in California, all edible. *F. chiloensis,* the Beach Strawberry, occurs on coastal beaches and nearby areas from San Luis Obispo Co. to Alaska and can be used as a ground cover. Two other species occur from Tulare Co. north.

Uses—Past and Present: Navajo Indians dried the fruit without changing the color or greatly reducing the size, but most foragers today cannot gather enough for preserving because they are too easily consumed in the field. There are many other kinds of wild fruit that are more abundant and less trouble to gather but none that have such appeal to the eye, smell, and taste as the strawberry.

The leaves, dried or fresh, are used to brew an excellent tea. The berries are a good source of thiamin, riboflavin, niacin, and Vitamin C. Upland game birds such as the California Quail feed on strawberry leaves and fruit, while the Catbird and Cedar Waxwing feed on the fruit only. The recipes for strawberries are legion, and I will give only one favorite.

STRAWBERRY SHERBET

8 oz buttermilk
1/2 tsp vanilla

1 cup whole straw-
  berries, cleaned
4 tsp sugar or honey

Place all ingredients in a blender and whip until thoroughly mixed. Store in freezer until firm. Transfer to refrigerator 30 minutes before serving to soften. Serves 1-2.

44. THIMBLEBERRY

Rubus parviflorus (pl. 3f) Family: Rosaceae

Description: Deciduous shrub, 1-2 m high, without prickles. The red-brown bark shreds in age. Leaves palmately 5-lobed. Showy flowers are scattered, white to pink. Petals 1.5-2 cm long. The flowers are larger than most species of *Rubus.* Fruit scarlet, in a hemisphere shape, 1-1.6 m broad. Blooms March through August with fruit following.

Distribution and Habitat: Open woods and canyons, especially near streams below 8000 ft (2500 m) in mountains of San Diego Co. through Sierra Nevada to Siskiyou and Humboldt counties. Coastal mountains from Santa Barbara Co. to Mendocino Co. Those of the coast are inferior in taste to the central and eastern berries.

Uses—Past and Present: Thimbleberries were used fresh and dried, pounded with meat and fat to form pemmican much as were blackberries, salmonberries, and blueberries. Tender young shoots can also be boiled or eaten fresh. For animal uses, refer to Blackberry. Recipes for Blackberry, Raspberry, and Strawberry may use thimbleberries.

THIMBLEBERRY SNOW

1 envelope unflavored gelatin	2 tsp lemon juice
	2 Tbsp sugar
1 cup thimbleberry juice	1 cup heavy cream
1 cup thimbleberries, chopped	

Soften unflavored gelatin in berry juice (made with about 1/3 cup sweetened crushed berries and water, then strained). Dissolve over boiling water and stir in berries. Chill until syrupy. Combine lemon juice, sugar, and heavy cream and whip until stiff. Fold gelatin mixture into whipped cream, pour into a mold, and chill for about 4 hours. Serves 4.

45. Mountain VIOLET
Viola purpurea (pl. 4a) Family: Violaceae

Description: Herb with basal leaves, simple, with smooth margins. Has 1-5 leaves, rounded, purple tinted, and somewhat succulent, 1.5-3 cm wide, 1.8-3.5 cm long on long petioles. Flower petals lemon-yellow with lower one in the shape of a spur or sac, 8-10 mm long; 2 upper petals purplish on the back; 3 lower with purplish-brown veins. Variable. Blooms April through June.

Distribution and Habitat: Dry slopes 1800-6000 ft (500-1800 m) with some varieties growing at higher altitudes. Pine forests, some in sagebrush areas. Mountains of San Diego Co. along western Sierra to near Mt. Lassen, Mt. Shasta, and Siskiyou Mountains; in coastal mountain ranges from Mt. Pinos to San Rafael Mountains and Lake Co. north. Twenty-four species of *Viola* occur in the state. All are edible but their value as a wildflower should be considered, since several species are on the rare and endangered plant list of the California Native Plant Society. Colors of the flowers range from white and yellow to lavender and dark purple. Some have heart-shaped leaves.

Uses—Past and Present: The crushed leaves of this wild relative to the pansy were reported in the seventeenth century to be used for boils and swellings. It has also been used ground up for many types of skin diseases. Wild violets are used in the South to thicken soups where they are called Wild Okra. Old-fashioned candied violets were a colonial favorite as was violet-flavored vinegar. Violet leaves make a tender, fresh green raw or cooked—an excellent source of vitamins A and C. Violet leaf tea is also a common forager drink. The seeds of violets are eaten by several different birds, such as Mourning Dove, California Quail, and junco. Rabbit and rodents also eat the seeds. Fritillary butterfly caterpillars are fond of the leaves. Most violets should be used as emergency food only; *V. tomentosa, V. halli,* and *V. cuneata* are rare and endangered species.

CANDIED VIOLETS

3 doz violet flowers   finely granulated sugar
1 egg white

Wash, drain, and dry violet flowers. Remove all green parts, keeping petals intact. Beat one egg white at room temperature until foamy but not stiff. Dip each flower in egg white, then in sugar. Place on waxed paper and separate petals with a toothpick. Dry in the sun or in a warm place.

VIOLET OMELETTE

1 doz large violet   salt and pepper
  leaves and blossoms   to taste
4 eggs   1 tsp butter
4 tsp light cream   or margarine

Wash violet leaves and blossoms, chill, and chop. Beat eggs until light yellow. Add cream, salt, and pepper. Melt butter or margarine in a skillet and pour egg mixture in. Lift edges to cook evenly. Sprinkle leaves and flowers on top of egg mixture. Fold omelette in half and cook until set. Serve hot. Serves 2.

46. Southern California Black WALNUT
Juglans californica (fig. 24) Family: Juglandaceae

Description: Familiar wild relative of the cultivated tree. Pinnate leaves with 11-15 leaflets, mostly 2-6 cm long, narrower than the cultivated. Fruit a round nut with green husk maturing to dark brown, develops in summer and fall.

Distribution and Habitat: Common below 2500 ft (800 m) in Valley and Foothill Woodland and canyons, Orange Co., Santa Ana Mountains, San Bernardino Co. to western Sierra, Ventura Co. Also occurs in Santa Lucia Mountains of Monterey Co. *J. hindsii* is found about old Indian campsites,

Fig. 24. WALNUT *(Juglans californica).* 3/4 X.

Lake and Napa counties to Contra Costa and Stanislaus counties.

Uses—Past and Present: Thick shells have contributed to this delicious wild nut's decline in popularity. Cahuillas used the hulls for dye in basketry. The stains caused by the hulls are an added deterrent to collecting today, but the stain-causing husks may be removed by drying or peeling with a knife. By wearing gloves the collector can avoid the stains altogether. Truly nutritious. Ground squirrels feed readily on them.

BLACK MAGIC PIE

1 cup chopped walnuts	1/4 cup melted butter
3 eggs	or margarine
3/4 cup milk or cream	3/4 cup brown sugar
3/4 cup dark corn syrup	1/2 cup self-rising flour
	or biscuit mix

Preheat oven to 350° (mod.). Shell nuts by letting husks rot off or by peeling them off while wearing gloves. Mix together eggs, milk or cream, corn syrup, and butter or margarine. Add brown sugar, self-rising flour or biscuit mix, and walnuts. Blend well; the mixture will be thin. Pour into a 10-in buttered pie pan and bake for 50 minutes or until it is puffed and set. Serve chilled with whipped cream. (This prize-winning pie makes its own crust.)

47. WATERCRESS

Nasturtium officinale formerly *Rorippa nasturtium-aquaticum* (fig. 25) Family: Cruciferae

Description: Prostrate aquatic with shiny green pinnate leaves. Familiar because of its cultivated uses. Has 5-15 leaflets. Flowers white with 4 petals, 3-4 mm long, 6 stamens. Blooms March through November.

Distribution and Habitat: Common in quiet water or slow streams, on wet banks below 8000 ft (2500 m) in many plant communities. A similar plant with yellow flowers has been used in the same way by campers and hikers.

Uses—Past and Present: Watercress is in constant demand in markets today for use in sandwiches, soups, and salads. Persians believed it would make children strong and it was much prized by the Moslems. Romans considered it excellent food for people with deranged minds. American Indians used the plant for liver and kidney trouble as well as to dissolve gallstones. If water in which it is gathered is polluted, the plant should be cooked or thoroughly rinsed in purified water. Best parts are those of the upper stems. Taste is typical of the Mustard family—peppery. To tame the distinctive taste, simmer it in two changes of water. Most hikers and backpackers find it delicious and refreshing as is.

Fig. 25. WATERCRESS
(Nasturtium officinale).
Actual size.

VIENNESE WATERCRESS SOUP

3 cups beef stock
2 medium potatoes,
  peeled and chopped
1 Tbsp salt
1 large onion,
  chopped

1/4 cup butter
  or margarine
1 egg yolk, beaten
1 cup milk or cream
6 cups chopped Watercress
1 Tbsp Moselle wine
  (optional)

Simmer beef stock (or 3 bouillon cubes in 3 cups water) with
potatoes, salt, and onion until tender. Add enough water to
keep the liquid level constant. Sauté Watercress in butter or
margarine and add to the potato stock. Simmer 10 minutes,
remove from heat, and stir in egg yolk and milk or cream, and
add wine. Serve hot or cold. Serves 6.

PLANTS TREATED ELSEWHERE WHICH MAY
ALSO OCCUR IN FOOTHILLS AND MOUNTAINS
(Refer to General Index for page numbers.)

Arrowhead
Beavertail Cactus
Bulrush
Canaigre
Cattail
Chia
Chickweed
Chicory
Fennel

Filaree
Horehound
Lamb's Quarters
Day Lily
Live Forever
Nasturtium
Nettle
Nut-Grass
Oat

Pinyon
Sagebrush
Sheep Sorrel
Shepherd's Purse
Sunflower
Milk Thistle
Sow Thistle
Yerba Santa
Yucca

DESERTS

48. Desert AGAVE
    (also Century Plant, Maguey) *Agave deserti*
    (fig. 26) Family: Agavaceae

Description: Fleshy-leaved perennial with prickles along edges of leaves. Leaves basal in a rosette, gray-green, with a whitish-gray coating, 1.5-4 dm long. Flowering stalk 3-5 m high. Flowers yellow with parts in multiples of three, 3.5-5 cm long, produced in inflorescences that resemble the hands of a banana. They sometimes require 20-40 years to mature, earning them the common name century plant. Vegetative reproduction of a new plant accompanies flowering, so the plant is not destroyed. Plants flower from May through July, then die.

Distribution and Habitat: Dry, rocky desert slopes in Colorado and Mojave deserts, 3000-5000 ft (900-1500 m), Creosote Bush Scrub, Shadscale Scrub. In Joshua Tree Woodland, *A. utahensis* is found, but has only 4 flowers per cluster. It is used in a similar manner.

Uses—Past and Present: Young flowering stalks were an important staple of desert Indian tribes. They were dug out of the basal leaf cluster with digging sticks and roasted for several days in earth pit ovens for a sweet-tasting starchy food which was eaten fresh or dried for future use as an important barter item. These same crowns were roasted by some tribes, such as the Apaches of the White Mountains and the Chiricahuas, for preparing an alcoholic drink. Today the young flowering stalks yield a liquid which is fermented into the drink known as pulque. When distilled, this is the chief ingredient of tequila.

The flowers were also eaten by the Indians. They were boiled or dried for winter use. The tough fibrous leaves were dried or pounded in water to yield fibers for ropes, bowstrings, brushes, and sewing materials. The spines with

99

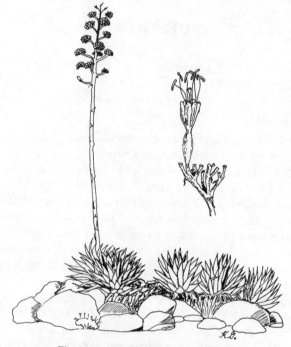

Fig. 26. Desert AGAVE *(Agave deserti).*

attached fibers were used as needles, while sometimes the entire young leaf was chewed as a tonic, though there are reports that raw agave is poisonous. The juice of the roots was sometimes applied to fresh wounds, and the seeds ground for flour. In southern California, the charcoal from agave was used by the Indian women for tattooing. The Agave Skipper Caterpillar was often roasted and easten as a delicacy. Agave serves as a food source to Band-tailed Pigeon and Bighorn Sheep. Its dried stalks are used by the Cactus Woodpecker for nesting sites.

A. shawii of San Diego Co. is on the rare and endangered plant list of the California Native Plant Society and should be used only in extreme emergencies. The conservation-minded forager will take care to use only limited amounts of the least-damaging portions of this elegant desert plant, regardless of species.

SAUTÉED AGAVE BUDS

1 cup agave buds   salt to taste
1 Tbsp butter or
  bacon drippings

Boil unopened buds in water for 5-10 minutes. Drain and cover with cold water for a few minutes. Drain and pat dry. Sauté in butter or bacon drippings and salt to taste.

Agave buds may also be prepared in a batter and fried tempura style.

49. BARREL CACTUS
(also Bisnaga) *Ferocactus acanthodes* (fig. 27)
Family: Cactaceae

Description: Globular or cylindric cactus with no branches. Height 2 m or more. The largest plants are approximately 20-30 years old. Spines are red to white, flowers yellow and funnel-shaped. Blooms appear in spring.

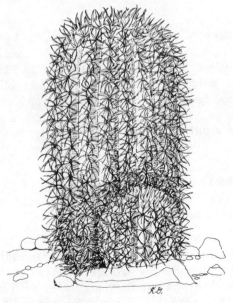

Fig. 27. BARREL CACTUS *(Ferocactus acanthodes)*.

Distribution and Habitat: Below 5000 ft (1500 m) in Creosote Bush Scrub, Joshua Tree Woodland, rocky slopes and walls mostly of the Mojave Desert.

Uses—Past and Present: Southwest Indians scooped the flesh out of these cacti and squeezed out the juice for emergency moisture as is taught in survival classes today. A novel use of the Barrel Cactus was as a form of oven. First the top of the cactus was sliced off, and the inside tissue scooped out. The skeleton of the cactus was then used for a container to bake other foods. This was accomplished by placing hot rocks in the cactus and replacing the lid on its top.

The spines were often used as awls. The Cahuillas removed the base of the buds and boiled them for food. The flesh of this and other species of cactus can be made into cactus candy, which is sold in stores in desert areas. It is my hope that you will use only cultivated cactus for these recipes, since their use destroys the beautiful plants. For recipes, see the *Cactus Cook Book* by Joyce Tate, cited in the references.

CACTUS CANDY

Obtain squares of cactus pulp from the cactus by slicing. Soak overnight in water, drain, and weigh. To each lb of pulp, add 1 lb of sugar and boil until clear. Cool and boil again the next day for 30 minutes. Repeat this cooling and boiling for 5 days. Lastly, boil until all the syrup is absorbed into the cactus, stirring constantly. If you desire colored candy, add a little fruit juice or vegetable dye to the first boiling.

50. BEAVERTAIL CACTUS
(also Prickly Pear, Tuna) *Opuntia basilaris* (fig. 28) Family: Cactaceae

Description: One of a large group of species with flat pads that resemble a beaver's tail. Stems are low and spreading, 1-3 dm long, often purplish. Flowers are orchid to rose in color. There is much hybridization among the many species of *Opuntia.* They are all edible and most are palatable, especially *O. occidentalis,* which actually bears the Prickly

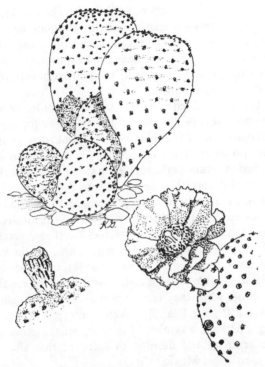

Fig. 28. BEAVERTAIL CACTUS *(Opuntia basilaris).*

Pear or Tuna common names. Other popular species are *O. ficus-indica, O. megacantha,* and *O. littoralis. O. basilaris* is historically the most cited for its Indian usages. Blooms March through June, with fruit following.

Distribution and Habitat: These cacti are common in the Mojave and Colorado deserts and in Creosote Bush Scrub and Joshua Tree Woodland below 4000 ft (1200 m). Some varieties grow in Valley Grassland, Chaparral, and urban areas where overgrazing has weakened the soil for support of less hardy species of plants.

Uses—Past and Present: Cacti of the flat-stemmed variety were a staple of the Indians of most western states. The entire plant was often used for food. Young fruits were gathered in the morning when the bristles were wet or rubbed in the sand to remove the tiny spines. These green tunas were then boiled or baked to the consistency of applesauce. The Blackfeet of the northern plains eliminated warts by rubbing the young spines into them. The Navajo picked the fruit of the Prickly Pear with great reverence. To appease the spirit they believed inhabited the plant, they offered the plant a hair from the gatherer's head in sacrifice.

Ripe fruits were dried and preserved for later use, while the padlike stems were peeled and used as a poultice for wounds and inflammations. These pads were also cooked much as the Mexicans do today. The young pad is flamed to remove the spines, then the stubs of the spines are sliced off with a sharp knife. The pad may then be sliced into "shoestrings" and substituted in green bean recipes, stews, and other dishes. I have removed the upper and lower skin and substituted the remaining tissue for eggplant in recipes for that vegetable.

In Mexico, the Prickly Pear is represented on the silver peso, the state flag, and the arms of the Republic. Mexican folklore states that in 1325 the Aztecs were being pursued by a hostile people when they came upon an eagle strangling a snake atop a Prickly Pear. The Aztecs interpreted this as a good sign, perhaps a symbol of their eventual victory over their adversaries, and decided to settle at that site—the present location of Mexico City.

The fresh fruits of many species are enjoyed today. They can be eaten right off the plant when ripe, or cooked when green, but care must be taken in handling to avoid the many sharp spines. The scooped-out watermelonlike insides are a real treat. The syrup made from mature fruits can be used to flavor candy, sauces, ice cream, and many other dishes calling for fruit. The spines and rinds are removed easily by placing the fruit in boiling water for a minute or two; they can then be easily peeled. The fruits are high in calcium, phosphorus, and vitamin C.

The fruits, seeds and stems serve as food for many different kinds of animals. Rodents, in particular, eat the seeds for food and often chew the pads for moisture. Sheep and deer browse on them also. Birds especially feed on the fruit, while the Cactus Wren makes its nest in the branches of certain species.

Many *Opuntias* are on the rare and endangered plant list for California and should not be picked.

BASIC PRICKLY PEAR SYRUP

1 qt Prickly Pear fruits   2 Tbsp lemon juice
water   3/4 cup sugar

Remove the spines and peel from the cactus fruit after boiling them a minute or two. Slice the fruit, discard any excess seeds, and add lemon juice and sugar to taste. Cook until mushy, drain off the juice and strain it through a fine strainer. Add more sugar (or honey if desired) and cook it down to a syrup. Use in making punch, toppings, pies, and so on. Makes 1 cup.

CACTUS JELLY

1 gal Prickly Pear fruits   10 Tbsp lemon juice
water   1 tsp cinnamon
5 cups sugar   (optional)

After removing spines, peel, wash and quarter a gallon or so of Prickly Pears or Tunas. Put in a kettle with barely enough water to cover. Cook slowly until fruit is tender and light in color (about 20 minutes). Run through a jelly bag or cheesecloth. Leave for several hours. Don't squeeze the bag or juice will be gummy. Skim off any particles that come to the top. Add 1 cup of sugar (or 2/3 cup honey) and 2-3 Tbsp lemon juice per cup of extract. Cook until syrup drops in sheets from spoon, about 30 minutes. Pour into hot, sterilized jars and seal. Cinnamon may be added for spiced jelly, about one tsp per 5 cups. Makes about 3 pints. Commercial pectin may also be used to jell the jelly in lieu of cooking. Follow the recipe for apple jelly.

DEEP FRIED CACTUS RINGS

4 young Beavertail Cactus   1 tsp celery salt
   stems   1/4 tsp pepper
1 tsp salt   1/2 cup milk

2 eggs   oil for deep frying
1 cup flour   catsup

Boil or steam cactus until tender, rinse in cold water, and slice shoestring style. Mix together until smooth salt, celery salt, pepper, milk, eggs, and flour. Put the two ends of each cactus shoestring together, forming a loop, and pierce with a flat, uncolored toothpick to hold the ends together. Dip cactus ring into batter then into preheated oil and fry until golden brown. Drain on a towel and serve while still warm with catsup. These may be done ahead and heated in the oven for 10 minutes prior to serving. For more tender rings, remove the fibrous strings just under the cactus skin before slicing. If the pads are young, this will not usually be necessary. To vary this recipe, add 3/4 cup shredded cheese or onion salt to the batter in place of salt. It is very popular as an appetizer with my friends.

51. BOXTHORN

(also Wolf Berry, Tomatillo, Squawberry) *Lycium fremontii* (pl. 4b) Family: Solanaceae

Description: Glandular, hairy shrub 1-3 m tall, with needle-like spines and spatulate leaves 1-2.5 cm long and 3-10 mm wide. Flowers are pale violet to lavender, funnel-shaped with the tube 1-1.5 cm long and lobes 1.5-3.5 mm long. Fruit is oval, red, and fleshy, 6-8 mm long. Blooms March to April, with fruit following.

Distribution and Habitat: Common in Colorado Desert below 6000 ft (1800 m). There are other species of *Lycium* that are found further north to Mono Co. They occur in Creosote Bush Scrub, Pinyon Juniper Woodland, Sagebrush Scrub, Chaparral, and Coastal Sage Scrub.

Uses—Past and Present: The abundance of the berries made them an important item in the diets of numerous Southwestern Indian tribes, particularly the Cahuillas, Zunis, and Hopis. The fruit was eaten fresh or dried in the sun like raisins. Dried berries were boiled into a mush or ground into flour and mixed with water. Today they are treated much like tomatoes which they resemble in color and flavor. Flavor varies from plant to plant, but the green berries should never be eaten. Gambel Quail eat the berries, as do Canada Geese, doves, and kangaroo rats.

LYCIUM SAUCE

1 onion, chopped
2 Tbsp butter or
   margarine, melted
2 tsp flour
2 cups fresh red
   Boxthorn berries

1/2 bay leaf
1/4 tsp salt
dash of pepper
dash of celery salt

Saute onion in butter, add flour, and blend. Add berries and bay leaf and cook 20 minutes. Remove bay leaf and add seasonings. Chill and use like catsup or serve hot with meals. Makes 1 cup. Lemon juice or vinegar may be added if a more tart sauce is desired.

52. CANAIGRE
(also Wild Rhubarb, Tanner's Dock, Pie Dock)
Rumex hymenosepalus (pl. 4c) Family: Polygonaceae

Description: Perennial herb with tuberous roots; stems somewhat reddish, 6-12 dm high. Oblong leaves very fleshy, 6-30 cm long with wavy margins. Leaves mostly basal. Fruit a triangled nutlet in pinkish clusters at top of main stem. Flowers January through May.

Distribution and Habitat: Common in dry, sandy places below 5000 ft (1500 m) in grasslands, Coastal Sage Scrub, Creosote Bush Scrub, Chaparral, Joshua Tree Woodland, from Baja California to Monterey Co.

Uses—Past and Present: The tuberous roots contain up to 35 percent tannin and have been used for tanning leather. In early spring, stems and leaf stalks are cooked like Rhubarb, which they quite resemble. Leaves are sometimes eaten, but are cooked in several changes of water to remove the bitter tannin. The Hopi Indians of Arizona treated colds with a tea of Wild Rhubarb root. The Navajos used the powdered root for treatment of sore throat, while others boiled it for the same purpose. A tea made from the root was sometimes used on skin sores. The Navajos also used the seeds in mush.

Other species of *Rumex* are often used in similar ways. The urban weed, *R. crispus,* Curly Dock, resembles Canaigre in appearance but does not have the tannin in such high concentrations. As a result, Curly Dock leaves make excellent potherbs and stuffings. Its stems when young and tender may also be used for pie or sauce. Another dock relative used in similar ways, but having a different appearance, is called Sheep Sorrel and is discussed in the Urban section.

MOCK RHUBARB PIE

4 1/2 cups tender
  Canaigre stems, cut
  into short pieces
1/4 cup flour
2 cups sugar

1 Tbsp butter
  or margarine
1 tsp cinnamon
dash of lemon juice
2-crust pie shell
  (unbaked 9")

Preheat oven to 375° (quick mod.). Take care to use only the youngest, most tender stems or your pie will be chewy. Steam or boil stems with just enough water to cover. When tender, drain and add remaining ingredients. Cook until thickened or stir in 2 Tbsp cornstarch that has been dissolved in a little water. Pour into pie shell, cover with top crust, and bake for 50 minutes. Cover top crust with foil if it starts to get too brown. Serve hot or cold. Top with vanilla ice cream if desired.

53. CATCLAW

(also Acacia) *Acacia greggii* (fig. 29) Family: Leguminosae

Description: Spreading deciduous shrub 1-2 m high. Branches armed with short, stout, curved spines. Leaves 2-5 cm long, twice compound into 4-6 pairs of leaflets which are 2-8 mm long. Flowers yellow, in cylindrical catkins 1-4 cm long. Fruit a pod, somewhat constricted between the seeds, 2-

Fig. 29. CATCLAW *(Acacia greggii)*. 3/4 X.

12 cm long and 1.5-1.8 cm across. In mid-August the light green pods turn reddish brown and make a fine show of color when abundant. Blooms April through June.

Distribution and Habitat: Common in stream beds and canyons below 6000 ft (1800 m), Creosote Bush Scrub to Pinyon-Juniper Woodland in Colorado and southern Mojave deserts. Occasionally in other shrubby areas of southern California.

Uses—Past and Present: It is reported that the extensive groves of Catclaw provided a substantial, though not preferred, source of food for the Cahuillas. Pods were eaten fresh or dried and ground into flour from which mush or cakes were prepared. The young pods are often bitter and require boiling to make them palatable. I have found that the seeds, when threshed out of the mature pods, make a good bean when added to ham dishes. They have a high protein and oil content.

Acacia flowers are a source of high-grade honey. A gum much like gum arabic exudes from the bark and is used locally in Mexico. The wood was used for construction and fires by the Cahuillas, while the highest branches are a favorite nesting site for verdin. Lower ones often are used for temporary shelter by pursued jack rabbits.

CATCLAW BEAN CHOWDER

1-2 cups dried Catclaw beans
1 cup diced ham
 or salt pork
1 small onion, diced
2 smoked ham hocks
dash of pepper

1 bay leaf or 2 dried
 juniper berries
1/2 cup diced carrots
1/2 cup diced potatoes
1/2 cup chopped celery

Cover beans with about 1 qt water and soak overnight. Discard water, recover beans with approximately same amount of water, add remaining ingredients, and bring to a boil. Cover and simmer until all vegetables are tender (about 2 hours). Add more water as needed. Serves 4-6.

54. CHIA
Salvia columbariae (pl. 4d) Family: Labiatae

Description: Annual, 1-5 dm tall with mostly basal leaves, finely pubescent and ovate in outline, 2-10 cm long, with many divisions. Blue flowers are in rounded clusters spaced along the stem. Flowers are 2-lipped and 12-16 mm long. Usually blooms March to June with mountain varieties continuing to bloom through summer. Seeds follow in 1-2 months and are best collected before fall winds.

Distribution and Habitat: Common below 4000 ft (1200 m) in open spaces of Coastal Sage Scrub, Chapparal, Foothill

Woodland and Creosote Bush Scrub throughout southern California and in inner coast ranges from Mendocino Co. south, and Sierra foothills from Calaveras Co. south.

Uses—Past and Present: Chia was one of the most important seed plants for the Indians of the West Coast. The seeds were gathered by beating them into baskets, then roasted or parched and ground into flour for mush or cakes. The seeds have a glutinous quality which causes them to swell when placed in water. Because of this they were sometimes placed under the eyelids to remove foreign matter. It is said that the nutritional value of Chia was such that one teaspoonful was sufficient to sustain an individual going on a forced march for twenty-four hours. In early California, Chia seed was used in a poultice for infections and was placed in alkaline water to make the water palatable.

Other species of *Salvia* are known as the sages and are aromatic plants used in cooking. *S. mellifera,* the Black Sage, for instance, is used for tea and was used by early settlers to season sausage, poultry, and meat stuffings.

A refreshing drink can be made of Chia seeds. The Spanish settlers used approximately one teaspoonful of seeds per drink plus a little sugar or lemon. Chia is sold in natural food stores and markets for its tender sprouts, which are delicious in salads. The seeds can also be ground to make muffins or breads. *S. columbariae* var. *ziegleri* from Riverside Co. is on the rare and endangered plant list and should not be harvested.

SOUR CREAM CHIA MUFFINS

6 Tbsp butter or margarine	1 1/4 cups Chia seeds (ground or whole)
1 1/2 cups brown sugar	1 1/2 cups flour (unbleached white or wheat)
1/2 tsp salt	
4 eggs	
1 3/4 cups sour cream	1 tsp baking soda
	1/4 tsp nutmeg

Preheat oven to 425° (hot). Cream together butter or margarine, brown sugar, and salt. Add eggs and sour cream.

In a separate bowl, combine Chia seeds, flour, baking soda, and nutmeg. Mix together wet and dry ingredients until just blended. Fill muffin cups 1/2 full and bake for 12-13 minutes. Variations: Sprinkle muffins with sugar or sugar and cinnamon before baking or add 1/2 cup raisins, currants, or floured blueberries to the batter. Makes 3 1/2 dozen.

55. CHUPAROSA

Beloperone californica (fig. 30) Family: Acanthaceae

Description: Shrub with ovate leaves 1-1.5 cm long, short petioled, and deciduous. Calyx of flower 4-5 mm long, corolla red, 3-3.5 cm long, irregular. Two stamens exceed length of upper lip of corolla. Fruit a capsule 1.5-2 cm long with yellowish-purple seeds. Flowers in angles of leaves, sometimes in fours. Blooms from March to June, but some blossoms may be present all year.

Distribution and Habitat: Common along watercourses below 2500 ft (800 m) in western and northern edges of Colorado Desert to Baja California.

Uses—Past and Present: The Papago Indians still eat the blossoms, which resemble cucumber in flavor. Linnets and sparrows bite off and eat the nectar-filled flower bases, while hummingbirds probe them for insects and nectar. The Diegueño Indians are known to have sucked the flower for its nectar, as desert hikers today do. Because of the limited abundance of this pretty flower, it is included here mainly for its possible survival value.

56. CREOSOTE BUSH

Larrea divaricata (fig. 31) Family: Zygophyllaceae

Description: A strong-scented evergreen shrub with opposite leaves. The leaves divide into sessile leaflet pairs. Flowers are yellow, with petals 5-8 mm long and twisted. Small fruit is covered with white hairs. Blooms April-May.

Distribution and Habitat: This is the dominant shrub over large areas of desert from southern California to Inyo and Kern counties up to about 5000 ft (1500 m).

Fig. 30. CHUPAROSA *Beloperone californica).* Actual size.

Fig. 31. CREOSOTE BUSH *(Larrea divaricata).* 1/2 X.

Uses—Past and Present: Perhaps because Creosote Bush is such a dominant shrub over large areas of the southwestern desert and dry slopes, it was frequently employed by the Indians for medicinal purposes. The Cahuillas made a tea from the stems and leaves, which was thought to cure a variety of ailments including colds, infections, and bowel complaints. The tea was also considered a decongestant and a general health tonic. It was often sweetened with honey for these purposes. To relieve congestion, modern Cahuillas boil leaves in a pot or bowl and cover their head with a blanket to inhale the steam.

Other remedies made from the bush included poultices for wounds and infections. Powdered, crushed leaves were applied to sores and wounds. The Pima Indians of Arizona used its sap for toothache. Creosote Bush was officially in the U.S. pharmacopoeia from 1842 through 1942 and was used as an expectorant and pulmonary antiseptic. Today a substance is obtained from the leaves and twigs which delays or prevents butter, oils, and fats from becoming rancid.

The Creosote Bush provides shelter and shade for many rodents, especially the kangaroo rat. Seeds are eaten by these animals also.

57. DESERT HOLLY
(also Saltbush) *Atriplex hymenelytra* (fig. 32)
Family: Chenopodiaceae

Description: Low shrub, white scales, 2-10 dm tall. Leaves alternate, rounded to rhombic with obtuse tip, 1.5-3.5 cm long, toothed. Flowers small, male and female. Seeds brown, about 2 mm long. Blooms January through April.

Distribution and Habitat: Dry alkaline areas, Creosote Bush Scrub and deserts to Utah.

Uses—Past and Present: The attractive silvery leaves of this plant make it a desirable Christmas decoration and much of it is solid for that purpose. Like most of the saltbushes, it is a valuable desert graze plant. Southwestern Indians such as Zunis, Pimas, and Cahuillas used these plants widely. The tender stems were boiled and eaten as a potherb, while the seeds were gathered, parched, then ground into meal. Zunis ground the roots and blossoms of *A. canescens,* moistened them with saliva, and used the mixture as a treatment for ant bites. The seeds of *A. lentiformis* have been found to contain considerable amounts of ash, protein, and oil. Boiling improves the taste of the plant, and I have found this and its numerous relatives to be good emergency rations. *A. tularensis* and *A. vallicola* of northern California are on the rare and endangered plant list of California and should be used only in extreme emergencies.

58. IRONWOOD
(also Tesota) *Olneya tesota* (fig. 33) Family:
Leguminosae

Description: Grayish tree with scaly bark and pairs of spines below the leaves. Leaves are compound with 8-24

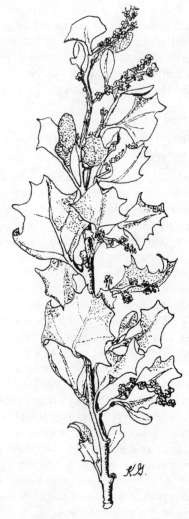

Fig. 32. DESERT HOLLY *(Atriplex hymenelytra).* Actual size.

leaflets. Purple flowers, 1 cm long, are scattered and appear before new leaf growth. Pod produced is 4-6 cm long with black seeds.

Distribution and Habitat: Below 2000 ft (600 m) in desert washes of the Colorado Desert.

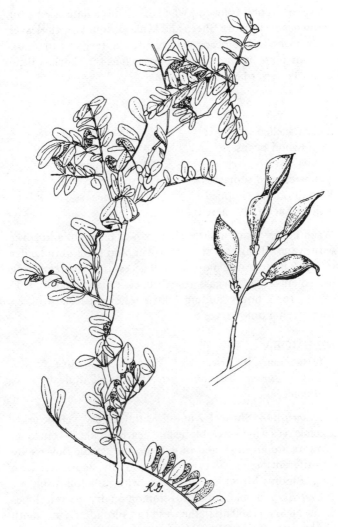

Fig. 33. IRONWOOD *(Olneya tesota)*. 3/4 X.

Uses—Past and Present: The seeds, which mature in late summer, are roasted and eaten for their peanutlike flavor. They are also ground into flour by the Cahuillas and used in gruels and cakes. I have enjoyed them boiled with ham. The extremely hard wood makes excellent tools and is also considered good firewood when dry. The Indians used it for arrowheads. Bighorn Sheep and Mule Deer browse the leaves of Ironwood. The beans may be used in recipes calling for navy or pinto beans or casseroles such as the Catclaw Bean Chowder described in this chapter.

INDIAN BEANS

2 cups shelled	1/3 cup sliced celery
Ironwood beans	3/4 tsp salt
6 cups water	1/4 tsp cinnamon
2 cups canned tomatoes	3 Tbsp brown sugar
1 small onion, chopped	1 Tbsp vinegar
4 slices bacon, quartered	

Wash beans. Place in large saucepan or bean pot and cover with water. Bring to a simmer, cover, and cook until tender (about 2 hours). Add more water if necessary. Add remaining ingredients. Place in casserole dish, cover, and bake at 350° (mod.) for 2 hours, adding a little water if needed to keep from drying out. Serves 6.

59. JOJOBA
(also Goat Nut, Deer Nut, Wild Hazel, Coffee Bush) *Simmondsia chinensis* (fig. 34) Family: Buxaceae

Description: Shrub 1-2 m tall, with leathery, yellow gray-green leaves which are oblong and nearly sessile, 2-4 cm long. Flowers are greenish and tiny. Male and female flowers are on different plants. Nuts are large and oily, acornlike, 2 cm long. Blooms March through May with nuts following.

Distribution and Habitat: Common on dry, barren slopes below 5000 ft (1500 m) in the western Colorado Desert south of the San Jacinto Mountains to San Diego and also in Chapparal in interior California, western Riverside and San Diego counties.

Uses—Past and Present: Nuts of the Jojoba were eaten without preparation by Indians and animals alike. The Cahuillas make a drink by grinding the nuts, boiling them,

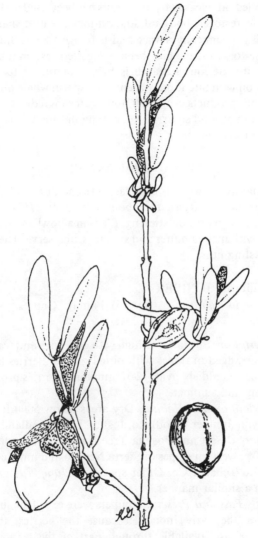

Fig. 34. JOJOBA *(Simmondsia chinensis).* Actual size.

and straining off the liquid. The Mexicans have a variation of this, producing a rich drink with roasted, ground nuts and egg yolk boiled with water, milk, sugar, and a vanilla bean.

The boiled oil released from the nuts has been used in the past as a hair restorer, body oil, and an ingredient in shampoo. The oil is chemically a wax and is indigestible to humans. Because of this, it has been suggested as an appetite depressant. Serious research is being conducted today on Jojoba oil with intentions of replacing sperm whale oil in the numerous products using that now-scarce ingredient. Sheep, goats, and ground squirrels are among the animals relishing the plant and its nuts.

ROASTED JOJOBA NUTS

Shell mature brown nuts and place on a cookie sheet. (Unripe nuts are bitter.) Roast for about 1 hour at 275°, stirring periodically to prevent sticking. Place in a bowl and sprinkle lightly with melted butter and salt, toss, and serve. These are mild-tasting nuts.

60. California JUNIPER
Juniperus californica (fig. 35) Family: Cupres-saceae

Description: Shrub 1-4 m high with gray bark and scalelike leaves arranged in threes with pitted glands. Berries bluish, changing to reddish, oval, 12-18 mm long. Berries produced in spring and summer.

Distribution and Habitat: Dry areas below 5000 ft (1500 m) Pinyon-Juniper Woodland, Joshua Tree Woodland, from Colorado Desert and Joshua Tree National Monument to Kern Co., western slopes of Sierra Nevada and inner coastal ranges to Tehama Co. Other species of *Juniperus* may be used in a similar manner.

Uses—Past and Present: Junipers were excellent survival food for the early Indians because the berries, though astringent, are available through part of the lean winter months. Some tribes preserved the berries by drying. They later ground the dried berries and baked them into cakes or made mush. Other groups roasted the berries, ground them, and made a beverage. Sometimes berries were used for

PLATE 1

a. Western BLUEBERRY
(Vaccinium occidentale) (p. 25)

b. BRACKEN FERN
(Pteridium aquilinum) (p. 26)

c. Sierra CHINQUAPIN
(Chrysolepis sempervirens) (p. 33)

d. Western CHOKECHERRY
(Prunus virginiana var. *demissa)* (p. 34)

e. Squaw CURRANT
(Ribes cereum) (p. 36)

f. Miner's DOGWOOD
(Cornus sessillis) (p. 38)

PLATE 2

a. Sierra GOOSEBERRY
   (Ribes roezlii) (p. 45)

c. California HUCKLEBERRY
   (Vaccinium ovatum) (p. 51)

e. Bigberry MANZANITA
   (Arctostaphylos glauca) (p. 55)

f. MINER'S LETTUCE
   (Calytonia perfoliata) (p. 59)

b. Mountain GRAPE
   (Mahonia pinnata) (p. 46)

d. Mariposa LILY
   (Calochortus nuttallii) (p. 54)

PLATE 3

a. White-Stemmed RASPBERRY
   (Rubus leucodermis) (p. 80)

b. WILD ROSE
   (Rosa californica) (p. 81)

c. SALAL *(Gaultheria shallon)* (p. 82)

d. SALMONBERRY
   (Rubus spectabilis) (p. 83)

e. SERVICE BERRY
   (Amelanchier pallida) (p. 83)

f. THIMBLEBERRY
   (Rubus parviflorus) (p. 92)

PLATE 4

a. Mountain VIOLET
   (Viola purpurea) (p. 93)

b. BOXTHORN
   (Lycium fremontii) (p. 106)

c. CANAIGRE
   (Rumex hymenosepalus) (p. 107)

d. CHIA
   (Salvia columbariae) (p. 110)

f. BULRUSH
   (Scirpus robustus) (p. 142)

e. SQUAW CABBAGE
   (Caulanthus inflatus) (p. 136)

ERRATA

Clarke, *Edible and Useful Plants of California*

Plate 4: Photo b should be captioned "CHIA *(Salvia columbariae)*"

Photo d should be captioned "BOXTHORN *(Lycium fremontii)*"

PLATE 5

a. Sweet FENNEL
(Foeniculum vulgare) (p. 150)

b. Feather Boa KELP
(Egregia laevigata) (p. 153)

d. SEA ROCKET
(Cakile maritima) (p. 172)

c. Stinging NETTLE
(Urtica dioica ssp. *holosericea)* (p. 160)

e. SEA WHIP
(Nereocystis leutkeana) (p. 172)

f. ALFALFA
(Medicago sativa) (p. 177)

PLATE 6

a. AMARANTH
(Amaranthus retroflexus) (p. 179)

b. Sugar BEET
(Beta vulgaris) (p. 184)

c. CHEESEWEED
(Malva parviflora) (p. 186)

d. LAMB'S QUARTERS
(Chenopodium album) (p. 198)

e. Black MUSTARD
(Brassica nigra) (p. 199)

f. PLANTAIN
(Plantago major) (p. 206)

PLATE 7

a. PURSLANE
   (Portulaca oleracea) (p. 208)

b. RADISH
   (Raphanus sativus) (p. 210)

c. Australian SALTBUSH
   (Atriplex semibaccata) (p. 211)

d. Milk THISTLE
   (Silybum marianum) (p. 220)

f. Strawberry GUAVA
   (Psidium cattleianum) (p. 240)

e. EUGENIA
   (Syzygium paniculatum) (p. 235)

PLATE 8

a. Day LILY
   (Hemerocallis aurantiaca) (p. 241)

b. Pot MARIGOLD
   (Calendula officinalis) (p. 243)

c. NATAL PLUM
   (Carissa macrocarpa) (p. 246)

d. PYRACANTHA
   (Pyracantha coccinea) (p. 247)

e. Poison Hemlock
   (Conium maculatum) (p. 150)

f. Death Camus
   (Zigadenus venonosus) (p. 28)

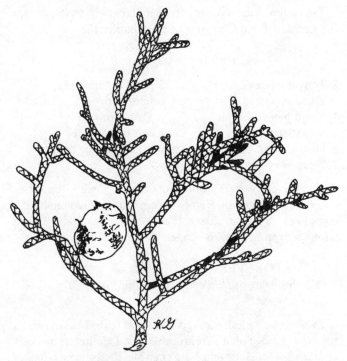

Fig. 35. California JUNIPER *(Juniperus californica)*. Actual size.

making a tea or simply chewed as a treatment for fevers and colds. Juniper berries are used commercially to impart a flavor to gin.

Juniper bark was used by the Indians in preparing medicine for treatment of colds, fever, and constipation. Some tribes used it for making clothing and mattresses, and in times of famine, ate the inner bark. A red dye was made from the ashes of the plant. The Navajos prized juniper for its clean-burning wood.

During June, Phainopeplas are seen in numbers in the juniper thickets eating the mistletoe berries that grow on them. Band-tailed Pigeons, quail, jays, and mockingbirds eat the fruit as do Coyote, fox, Opossum, and Ring-Tailed Cat. Antelope, deer, and Mountain Sheep browse on the foliage.

I use a few dried juniper berries to flavor meats. (See recipe for Catclaw Bean Chowder in this chapter.)

ROUND STEAK JUNIPER

2 lbs round steak, sliced or cubed	3 juniper berries, lightly crushed
1/2 cup melted butter or margarine	1 Tbsp lemon juice
3 Tbsp minced onion	1 cup white wine
1/2 cup chopped chives	salt and pepper to taste

Preheat oven to 400° (mod. hot). Sauté steak in butter or margarine with onion. Remove from heat and add remaining ingredients. Place in casserole dish, cover, and bake for 45 minutes or until steak is tender. Serve with rice. Serves 6.

61. LIVE FOREVER
(also Stonecrop) *Dudleya saxosa* (fig. 36) Family: Crassulaceae

Description: Fleshy-leaved perennial herb. Leaves form a rosette at base of plant, are lanceolate and semiterete in cross-section, 3-10 cm long and 1-1.5 cm wide. Fleshy stems slightly reddish with yellow flowers about 10-12 mm long, turning reddish in age. Flowers appear April through June.

Distribution and Habitat: Dry, rocky areas below 7000 ft (2100 m), Creosote Bush Scrub, Chaparral, Pinyon-Juniper Woodland, mostly desert mountains of San Bernardino Co. and slopes of San Jacinto and Laguna mountains, Panamint Mountains, Inyo Co. Other species occur in many areas throughout the state. *D. edulis,* a white-flowered species occurring in the southern part of the state, is known as Mission Lettuce for its use in salads.

Uses—Past and Present: These plants were considered a delicacy by the Cahuilla Indians. The fleshy leaves are eaten raw. Stems are slightly sweet and are refreshing to chew, though often leaving a chalky aftertaste. Many *Dudleyas* are on the rare and endangered plant list, so it is included here chiefly for its survival value.

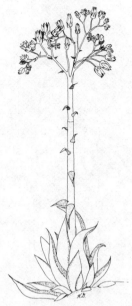

Fig. 36. LIVE FOREVER *(Dudleya saxosa).* 1/4 X.

62. MESQUITE
(also Honey Mesquite) *Prosopis glandulosa* var.
torreyana formerly *P. juliflora* (fig. 37) Family:
Leguminosae

Description: Low tree or large shrub with several trunks, 3-
7 m high. Deciduous compound leaves are bright green with
7-17 pairs of leaflets 1.5-2.3 cm long. Flowers are small and
greenish in slender spikes. The yellow pods are flat and occur
in one to several pods in a cluster. They are 0.5-1.5 dm long
and constricted between the seeds. Blooms April through
June, pods ripening in fall.

Distribution and Habitat: Common in washes and low
ground below 3000 ft (900 m), Creosote Bush Scrub and
alkali regions of Colorado and Mojave deserts. A similar
species, *P. pubescens,* is known as the Screw Bean or Tornillo
because the legumes or pods coil many times forming a
springlike fruit.

Fig. 37. MESQUITE *(Prosopis glandulosa* var. *torreyana).* 1/2 X.

Uses—Past and Present: Yumans, Mohaves, Cocopas, Pimas, and Papago Indians of Arizona, as well as other desert tribes agree this tree was one of their most important food sources. All parts were used in some way; the beans were even bartered. The long, yellow, pleasantly sour bean pods were collected and dried. Generally, the entire pod with seeds was ground into flour and then made into gruel or bread. Large quantities were eaten fresh or ground up and used for a drink. The Cahuillas of southern California gathered the blossoms as well as the pods. Blossoms were roasted and shaped into balls for ready eating or made into tea. Pods were picked both green and mature and ground into meal which was dampened with water and allowed to harden into cakes. Pieces were then broken from these cakes and eaten or made into a mush or beverage.

In addition to its uses as a food, the trunk of the Mesquite was often used in making mortars, while smaller limbs were made into furniture as well as digging sticks, bows, and arrows. The thorns were used in puncturing the skin for tattoos. The sap of the tree was used for a snack, for glue, or as a wash for wounds and sores. Boiled sap was mixed with mud and packed on the head to remove lice. A side effect of this treatment was that it dyed the hair black. Fibers from the roots and bark were used in making baskets, while a tea from the leaves was frequently used as an eyewash or drunk for headache and stomach trouble. Large game such as deer, antelope, and mountain sheep eat Mesquite pods. Rabbits, wood rats, quail, and squirrels gather near Mesquite for protection, food, and shelter.

MESQUITE PUNCH

4 cups dried	pinch of cinnamon
Mesquite pods	dash of ground cloves
1 Tbsp brown sugar	

Wash and break up Mesquite pods. Cover with water and boil 2 hours, adding water if necessary. Mash frequently. Reserving liquid, wring and break up pods by hand or put through a blender or grinder. Return to liquid and simmer, tightly covered, for 1/2 hour. Strain off liquid. To each cup of liquid, add 1 Tbsp brown sugar, a pinch of cinnamon, and a sprinkle of ground cloves. Heat and stir until sugar is dissolved. Serve warm or chilled. Serves 4.

63. Nevada MORMON TEA
    (also Squaw Tea, Mexican Tea, Miner's Tea,
    Joint Fir) *Ephedra nevadensis* (fig. 38) Family:
    Ephedraceae

Description: Broomlike, seemingly leafless shrub about 1.2 m high. Jointed stems are pale green when young, with a white bloom, turning yellow or gray with age. Scalelike papery leaves are very small and occur in pairs which fall off, leaving gray bases. Fruiting structures are small male cones 4–8 mm long, yellow to light brown with protruding stamens;

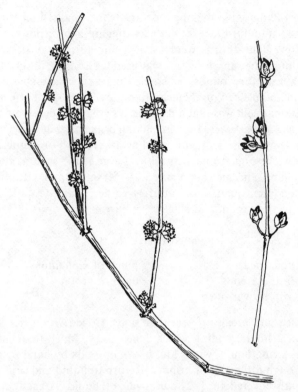

Fig. 38. Nevada MORMON TEA *(Ephedra nevadensis).* 3/4 X.

female cone is light brown to yellow-green and consists of small scales. Blooms March to April.

Distribution and Habitat: There are seven species of *Ephedra* in California and all inhabit warm, arid regions. Common on dry slopes and hills below 4500 ft (1400 m) in Creosote Bush Scrub, Joshua Tree Woodland, the Mojave Desert to Owens Valley.

Uses—Past and Present: Travelers, old and new, brew a tea by steeping the green or dry twigs in boiling water. The tea was considered a tonic for kidney ailments and stomach disorders as well as a blood purifier by Southwestern Indian tribes. The stems were chewed to relieve thirst, while the seeds were sometimes roasted and ground into flour to make a bitter bread or mush. The Pima Indians of Arizona dried the roots, ground them into flour, and sprinkled it on sores. Navajos boiled the twigs and leaves of *E. viridis* to produce a light brown dye. The drug ephedrine is obtained from a Chinese species. Quail, Mountain Sheep, and ground squirrels prize the seeds, and the entire plant is sometimes browsed by deer and sheep. *E. funerea* of San Bernardino and Inyo counties is on the California Native Plant Society's rare and endangered plant list and should be respected.

MORMON TEA

Bring water to a boil and place a handful of twigs for each cup desired in the water. Remove from the fire and let steep for 20 minutes or until desired strength is reached. Some prefer this flavored with honey and lemon.

64. OCOTILLO
(also Candlewood, Coach Whip) *Fouquieria splendens* (fig. 39) Family: Fouquieriaceae

Description: Resinous, spiny, whiplike shrub with stout stems all arising from the base of the plant and seldom branched. Grows 2-7 m tall. Loses its leaves sometimes several times per year, with each drought period. Flowers are bright red and occur at tips of branches. Blooms normally from April to May.

Distribution and Habitat: In dry, rocky places below 2500 ft (800 m) Creosote Bush Scrub of Mojave and Colorado deserts.

Uses—Past and Present: The woody stems are often used in Mexico today for living fences; cut stems are planted in the ground where they will reroot if watered. The stems may also be combined with adobe for building homes. The wax coating on the stems can be made into a leather dressing. The powdered root was used to treat swellings by the Apaches of New Mexico, while bathing in this decoction was used to relieve fatigue. The red tubular flowers may be soaked in equal amounts of water overnight for a beverage. This may be mixed with other fruit juices or consumed as is. Hummingbirds drink from the flowers.

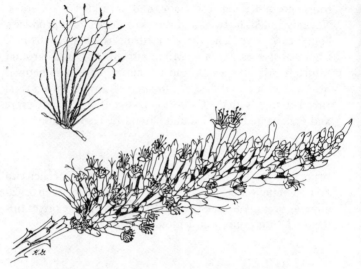

Fig. 39. OCOTILLO *(Fouquieria splendens).* 2/3 X.

65. Fan PALM

(also Desert Palm, California Fan Palm) *Washington filifera* (fig. 40) Family: Palmae

Description: Tree 10-15 m high with trunk 6-10 dm thick. Leaves gray-green, petioles 1-2 m long and about 5 cm wide. Blades 1-2 m long with 40-60 folds, segmented almost to the middle. This is the familiar palm with a fan-shaped leaf. Flowers in clusters at top of tree, whitish, each about 8 mm long. Fruit hard, 8-10 mm in diameter, with 1 brown seed. Blooms in June, fruit following.

Distribution and Habitat: In groves in moist alkaline spots, especially seeps, springs, and streams below 3500 ft (1100 m) on western and northern edges of Colorado Desert and into Mojave Desert. Naturalized in Kern Co.

Uses—Past and Present: The Cahuilla Indians made extensive use of the fruit of this, our only native, palm. The thin, sweetish pulp is still eaten fresh and the seeds ground into meal for making bread and porridge. Some fruits were dried in the sun and stored in ollas for future use. These were

Fig. 40. Fan PALM *(Washingtonia filifera)*.

then ground into flour that included both flesh and seed. A beverage was made by soaking the fruits in water. At least one report states that jelly was made from the datelike fruit. Seeds were considered excellent material for filling gourd rattles, while the petiole of the leaf was used to make cooking utensils. The leaves were used for roofs, baskets, and sandals, while the fiber was used in making cord. Parts of the palm were also used in the bow and drill used for making fire. The heart of the palm was sometimes boiled as an emergency food. Hooded Orioles use the fibers to construct their unique, pendant nests in the trees.

66. PALO VERDE
Cercidium floridum (fig. 41) Family: Leguminosae

Description: Shrub or small tree with distinctive green
bark. It reaches up to 10 m in height. Leafless most of the
year, leaves 1-2 cm long with 1-3 pairs of compound leaflets.
Frequently spiny. Flowers yellow, 8-10 mm long with 5 petals
and 10 stamens. Fruit a pod, 1-3 seeds and 4-8 cm long, often
somewhat constricted between the seeds. Seeds olive and
brown, 8-10 mm long, 6-7 mm wide. Blooms March through
May, fruit following.

Distribution and Habitat: Stream beds and low sandy
places below 1200 ft (400 m) Creosote Bush Scrub, Colorado
desert and Sonora Desert.

Uses—Past and Present: Cahuilla Indians gound the seeds
into meal, which was used in mush or cakes. It is not clear
whether the pods were ever used. A similar species with more
leaflets, *C. microphyllum* was used in the same way.

These plants are included here chiefly because they occur in
an area often devoid of emergency food.

67. One-leaf PINYON PINE
(also Nut Pine) *Pinus monophylla* (fig. 42)
Family: Pinaceae

Description: A small tree 5-15 m tall, usually with a divided
trunk and a rounded or flat top in age. Needles occur singly,
are rigid and pale gray-green, 2.5-3.5 cm long. The cones are
3.5-5.5 cm long with 4-sided scales. The nuts ripen in summer
with crops being produced only every other year under ideal
conditions.

Distribution and Habitat: Dry, rocky places 3500-9000 ft
(1100-2800 m) in Pinyon-Juniper Woodland and scattered on
the base of the Sierra, both sides from Mono Co. south. Two
other species occur mostly on the east side and bordering
deserts: *P. edulis* has 2-3 needles per cluster and *P.
quadrifolia* has 4 needles per cluster. *P. edulis* is listed on the
rare and endangered plant list of the California Native Plant
Society and is not recommended for use.

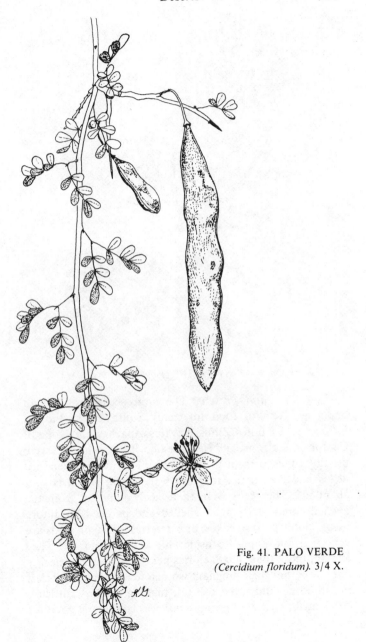

Fig. 41. PALO VERDE
(Cercidium floridum). 3/4 X.

Fig. 42. One-leaf PINYON PINE *(Pinus monophylla)*. 1/2 X.

Uses—Past and Present: The large seeds provided an important item of food for many Southwestern Indians including the Paiutes, Zuñis, Navajos, Apaches, and Hopis of California, Arizona, and New Mexico. The green cones were usually knocked from the trees with long poles, piled into heaps and set on fire to remove the pitch, which is very abundant. The seeds were then easily removed from the opened cones. They were shelled and parched for future usage, pounded into cakes, or eaten fresh. Pinyon seeds are very rich and supply considerable protein and fat. They contain more than 3000 calories per pound. Some Apaches did not permit their pregnant women to eat pinyon nuts lest the unborn child grow too fat, making delivery difficult. After birth, a gruel of ground nuts was considered good for the child. Seeds were sometimes obtained by robbing pack rat nests, where they would be cached in large quantities. Pinyon and other jays eat quantities of the seeds. Sheep and bears browse the plant.

The gum of the tree was used for chewing to ease sore throat, applied to burns and sores, and used as a glue. The Hopis and Apaches used it to repair and waterproof pots and baskets. The inner bark, though tough, saved many tribes from starvation during poor years.

Today, foragers still visit the trees to harvest the nuts, which are the largest of the pine nuts. Green cones may be roasted in an oven or over a grill for about an hour at 350° to open them up. Place a pan or aluminum foil in the bottom of the oven to catch the pitch. Pinyon nuts are easy to shell while they are still warm: roll vigorously with a rolling pin between two damp cloths. The nuts are delicious raw or mixed with vegetables such as green beans or peas. They may also be ground into a meal or flour for use in baked products.

PINYON MUFFINS

1 cup ground pinyon nuts	1/2 cup milk
	3 Tbsp honey
1/2 cup flour	1/2 tsp salt
2 tsp baking powder	

Preheat oven to 350° (mod.). Combine dry ingredients and then add wet ingredients and mix until just blended. Fill greased muffin cups 1/2 full and bake 30 minutes. Makes 10.

PINYON COOKIES

1/2 cup butter or margarine, softened	1 tsp grated lemon peel
	1 tsp lemon juice
2/3 cup light brown sugar	2 3/4 cups flour
	1/2 cup pinyon nuts
3 egg yolks	3 Tbsp honey, heated

Preheat oven to 325° (slow mod.). Cream butter or margarine with sugar. Blend in egg yolks. Add lemon peel

and juice and stir in flour until well blended. Measure about 1
Tbsp dough for each cookie and press into desired shape on a
greased cookie sheet. Sprinkle pinyon nuts over cookies and
press in firmly. Brush honey over cookies. Bake for 15
minutes or until golden. Makes about 3 1/2 dozen.

68. Big SAGEBRUSH
(also Great Basin Sagebrush) *Artemisia tridentata*
(fig. 43) Family: Compositae

Description: Aromatic, rounded evergreen shrubs, 0.5-3 m
high with a silvery leaf and a familiar sage odor. Many of the
leaves have 3 teeth on the tip, thus the specific name. Leaves
are 1-4 cm long and 0.2-1.3 cm wide. The corolla is funnel-
shaped, 5-toothed, and 2-3 mm long. Blooms August to
October.

Distribution and Habitat: Desert areas 4000-9500 ft (1200-
3000 m), Coastal Sage Scrub, Pinyon-Juniper Woodland,
getting up into Yellow Pine Forest throughout the Sierra
Nevada; from San Diego Co. through edge of deserts along
Sierra Nevada then to Modoc and Siskiyou counties north.
There are many species of this Sage which belong to the
Sunflower family. These are not the true Sages, however,
which are in the genus *Salvia* and belong to the Mint Family.

Uses—Past and Present: Cahuilla women gathered large
quantities of seeds, which they parched and ground to make
flour. They made a bitter tea from the leaves to use as a
treatment for sore eyes and colds, as a hair tonic, and to
alleviate stomach disorders. The fragrance of the wood
makes this sage a popular barbecue wood. Another species,
A. californica, was one of the most important medicinal
plants to the Cahuillas. Products made from the plant were
considered essential for proper maturation of girls into
women. Sagebrush provides essential cover for small desert
animals and is a favorite nesting site for Gray Vireo, Black-
throated Sparrow, California Sage Sparrow, and Costa's
Hummingbird. Mule Deer eat the foliage and rodents eat the
seeds.

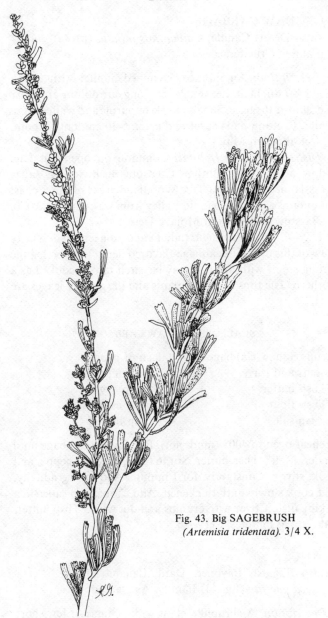

Fig. 43. Big SAGEBRUSH
(Artemisia tridentata). 3/4 X.

69. SQUAW CABBAGE
(also Desert Candle) *Caulanthus inflatus* (pl. 4e)
Family: Cruciferae

Description: An inflated, stemmed annual with simple
stems 2-7 dm high. Leaves, 2-7 cm long, are oblong and clasp
the stem at the base. Sepals purple or purple and white; petals
white. Fruiting pods stout and erect, 5-10 cm long. Blooms
March through May.

Distribution and Habitat: Common on open flat land
below 5000 ft (1500 m) in Creosote Bush Scrub, Valley
Grassland, and Joshua Tree Woodland areas of inner coast
mountains and San Joaquin Valley from western Fresno Co.
to Barstow and western Mojave Desert.

Uses—Past and Present: Indians and pioneers made a tasty
stew of this delicious cabbage-flavored herb. They boiled the
young stems with meat. It may be eaten raw or cooked as a
potherb. The tops of young plants and the greener leaves are
best.

SCALLOPED SQUAW CABBAGE

5 cups Squaw Cabbage	dash of pepper
stems and leaves	2 cups milk
3 Tbsp butter	1/2 cup dry
3 Tbsp flour	bread crumbs
1/2 tsp salt	3 Tbsp butter

Preheat oven to 400° (mod. hot). Boil Squaw Cabbage until
tender. Melt 3 Tbsp butter. Stir in flour, salt, and pepper and
cook, stirring constantly, for 1 minute. Add milk gradually,
and cook slowly until thickened. Add cabbage to sauce in a
baking dish. Cover with crumbs and dot with 3 Tbsp butter.
Bake 15-20 minutes. Serves 6.

70. YUCCA
(also Spanish Bayonet, Datil, Banana Yucca)
Yucca baccata (fig. 44) Family: Agavaceae

Description: A shrublike plant with a single or few short
stems surrounded by narrow stiff leaves at the expanded

base. This species of *Yucca* is distinguished by its grayish-blue or green leaves. It sits flush with the ground while others may have a trunk. Flowers on a central stalk, with 3 fleshy petals 1-1.5 dm long and 3 sepals 2.5-4 cm wide. Flower clusters are fleshy with red-purple tinges on the otherwise cream-colored parts. Fruit, which follows the flower, is approximately 17 cm long, fleshy, and resembles a short banana in shape. Blooms May-June.

Distribution and Habitat: Found on dry slopes 3000-6000 ft (900-1800 m) in Joshua Tree Woodland, Chapparal, Creosote Bush Scrub, and mountains of southern California. Other species of *Yucca* which are suitable for eating are found in northern California mountains and southern deserts. *Y. schidigera,* Spanish Dagger, was used extensively. The Joshua Tree is *Y. brevifolia.*

Uses—Past and Present: All parts of the southern *Yuccas* were used by the Indians to some extent. The tough, fibrous

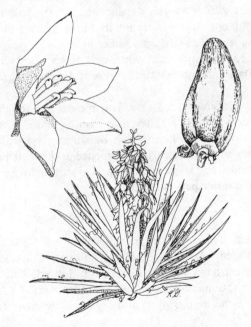

Fig. 44. YUCCA *(Yucca baccata)*. 1/2 X.

leaves yielded an important cordage for ropes, nets, hats, shoes, and mattresses, and sometimes paintbrushes were made by fringing them. The leaves were pounded in water to release the fibers. During World War I, *Yucca* fiber from Texas and New Mexico was used to make eighty million pounds of bagging and burlap. The ripened fruit was eaten raw or roasted, usually with the bitter outer skin removed. It was sometimes worked into cakes and dried for future use.

I have used the young flowers in soup and find them to be quite tasty when the central green heart (actually the ovary) is removed. Older flowers of some species are too bitter. The seeds of some species were ground into flour. The roots were cleaned and pounded to be used for soap by some tribes. This was called "amole" and was used to wash hair during wedding rituals and the naming of infants. The red roots of the Joshua Tree were used in basketry. Occasionally, the young flowering stalks of the *Yuccas* were cut from the plant and roasted in the same manner as the *Agave*. The night lizard, *Xantusia vigilis,* is entirely dependent upon the Joshua Tree and could not survive without it. Many species of birds, including the Red-shafted Flicker, Downy Woodpecker, Flycatcher, titmouse, and wren nest in the Joshua Tree. Woodrats sometimes climb the trees and gnaw on leaves or fruit.

Try *Yucca* flower petals tossed in omelettes, or add them to tomato or onion soup. They may also be added to tossed salads or deep fried like squash blossoms. The fruit may sometimes be gathered early and ripened at home. Ripe fruit may be seeded and sliced and used as a substitute for apples in a pie if the outer peel is removed.

YUCCA FRUIT SNACK

Boil fruit of *Yucca baccata* for 20-30 minutes. Drain, cool, peel, and seed. Mash pulp and return to pan. Cook until of desired consistency for jam. Sweeten if desired. Thickened with flour, this makes a good filling for turnovers. To dry for a snack, spread in thin layers and dry in a slow oven. Roll or fold.

PLANTS TREATED ELSEWHERE WHICH
MAY ALSO OCCUR IN DESERTS
(Refer to General Index for page numbers.)

Brodiaea
Currant
Filaree
Mariposa Lily
Milkweed
Tansy Mustard

Onion
Peppergrass
Russian Thistle
Shepherd's Purse
Sunflower

WETLANDS
Ponds, Streams, Marshes, Seashore

71. ARROWHEAD
(also Wappato, Tule-Potato) *Sagittaria latifolia*
(fig. 45) Family: Alismataceae

Description: Perennial aquatic herb with milky juice. The leaf blades are arrow-shaped, 1.5-6 dm long. Flowers conspicuous and in whorls of 3 with white petals 1-2 cm long and with many stamens. Quite variable. Bears tubers on roots about the size of a walnut. Blooms July through August.

Distribution and Habitat: One of five species bearing edible tubers found in slow streams or ponds, meadows or marshes throughout the state below 7500 ft (2300 m).

Uses—Past and Present: Arrowhead was reportedly eaten by Lewis and Clark on their expedition to the Northwest. Both Indians and Chinese are known to have used Arrowhead. In late summer, Indians loosened the tubers with their toes, floated them to the surface, and placed them in baskets. Then they baked, skinned, and ate them whole or mashed, and sometimes raw. They sometimes dried the roots for later use, or roasted and ground them for flour.

Pulling the plant from the water will only leave the tuber in the mud, since they are often several feet from the plant. A rake or hooked pitchfork can be used to loosen the tubers, and those which escape you will help spread the plant for next year's crop. Raw tubers are bitter, but boiled, fried, roasted, or creamed they taste much like potatoes. They may be substituted in any potato recipe—baked with sour cream and chives, boiled and mashed with butter and milk, baked with cheese, made into cold mock potato salad, added to stew, or served with meat. Tubers of *S. latifolia* are often too large or too deeply buried to be useful to waterfowl, but the small flattish seeds are eaten by ducks, and available tubers are valuable forage to wildlife.

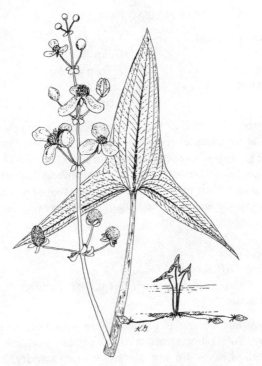

Fig. 45. ARROWHEAD *(Sagittaria latifolia)*. 1/2 X.

WAPPATO SALAD

1 qt Arrowhead
  tubers
4 eggs
1 cup sweet pickles
1 medium onion
1 Tbsp pickle juice

1 cup mayonnaise or
  salad dressing
1 Tbsp prepared mustard
1 Tbsp chopped parsley
salt and pepper to taste
wine vinegar and
  paprika (optional)

Boil and peel the tubers and eggs. Dice tubers, eggs, pickles, and onion and mix with remaining ingredients. A more sweet-sour taste can be created by adding wine vinegar to taste. Sprinkle with paprika. Serves 6-8.

ARROWHEAD PANCAKES

3 cups Arrowhead tubers	1 tsp salt
juice of 2 lemons in a bowl of water	1 cup flour
	2 Tbsp oil or melted butter
1 egg	1 tsp sugar

Rinse tubers and peel as thinly as possible so as not to waste the edible core. Put tubers in bowl of water and lemon juice as they are peeled—this prevents discoloration. When all tubers are peeled, discard lemon water, and grate tubers over a bowl, retaining juice. Stir in remaining ingredients. Spoon onto hot greased griddle and serve with maple syrup or other favorite pancake accompaniments.

72. Prairie BULRUSH
(also Tule) *Scirpus robustus* (pl. 4f) Family: Cyperaceae

Description: Perennial, grasslike herb with tuber-forming rhizomes. Single long stems, triangular in cross-section, 0.5-1.5 m tall. Leaves 4-6 mm wide, sometimes wider. Scaly flowers are reddish-brown to straw-colored and produced at the very tip of the stem. Blooms April through August.

Distribution and Habitat: This and larger species of Tule occur commonly in Freshwater Marsh, Coastal Salt Marsh, Alkali Sink, and throughout the state in wet ground. Commonly found where cattails grow.

Uses—Past and Present: Indians favored the sweet roots of Bulrush for grinding into flour. The Cahuillas gathered and ate the seeds raw or made them into mush. (Recent reports indicate raw, unprocessed seeds may be too prickly, however.) The Indians also made cakes of the pollen, and used the stalks to weave bedding, matting, and roofing materials. Bulrushes are sweeter than cattails, generally, and may be used in the same ways. The leading shoots that will be next year's plant stems are good raw or cooked. The roots may be crushed in water, strained of fibers, and made into flour. Seeds, ground into meal and combined with a like amount of other flour, can be used in muffins and bread products. The base of the plant is tender and edible raw or boiled.

The hard-coated seeds (akenes) are one of the most common foods of ducks, marsh birds, and songbirds who frequent the wetlands, specifically: Baldpate, Bufflehead, Mallard, Pintail, Shoveller, Blue-winged and Cinnamon Teals, Greater and Lesser Scaup, Avocet, Marbled Godwit, Clapper, Virginia and Sora Rail, Long-billed Dowitcher, and Tricolored Blackbird. Stems and underground parts are frequently eaten by Canada Geese and White-fronted Geese. The dense bulrush cover provides important nesting areas for waterfowl, marsh wrens, and blackbirds, and protection for raccoons and muskrats.

BULRUSH FLOUR

Clean roots and dry in sun or slow oven. Grind and remove fibers by straining and sifting. Pound remaining pulp into a flour (or use a blender). Another method is to boil peeled roots into a mush, strain out fibers, and use wet or evaporate the water. The flour is sweet.

TULE SHOOT HASH

1 lb ground beef
(for spicier hash,
use half sausage)
1 onion, diced
2 large tomatoes, diced

1/2 cup catsup
3 cups tender tule
shoots, diced
salt and pepper
to taste

Fry onion with ground beef. Add tomatoes, catsup, and tule shoots and simmer until the shoots are tender. Salt and pepper to taste. Serves 4-6.

73. Common CATTAIL
(also Soft Flag) *Typha latifolia* (fig. 46) Family: Typhaceae

Description: Tall perennial herb with creeping rhizomes. Unbranched stems are usually submerged at the base. Plant is

1-2.5 m high, leaves long and strap-shaped, 8-15 mm wide, nearly flat. Flowering spike is a familiar addition to dry plant arrangements. The male half produces pollen at the top, while the female portion produces the familiar tufts of feathery seeds which make up the lower half of the brown spike. Pollen and spikes mature early in summer.

Distribution and Habitat: Cattails are familiar inhabitants of subalkaline waters, Freshwater Marsh, and drainage ditches throughout the state below 5000 ft (1500 m). There are four species recognized in the state, all of which are useful.

Uses—Past and Present: The Cahuilla Indians gathered the underground parts in summer, dried them, and ground them into meal. They used the nutritious pollen to make cakes and mush, and the stalks for matting and bedding material and in constructing ceremonial bundles. Other tribes ate the young shoots and used leaves and leafsheaths for caulking material. The Paiutes of Nevada and Mono counties ate the flower stalks before the pollen was produced by boiling them or eating them fresh. Cattail pollen is very sacred to the Apaches of New Mexico who used it in puberty rites for girls. The Hopis of Arizona mixed the brown spike fuzz with tallow to make a chewing gum. Torches are made by cutting off the stalk and dipping the spike in coal oil. The fluff from the spike was also used as insulation, to line cradle boards, and for tinder when starting fires. The leaves have long served as material for weaving rush seating and baskets. The thousands of tiny, wind-distributed seeds are considered too small and hairy to be important food for wetland birds. The starchy cattail rootstocks are more important as a food source. White-fronted and Canada Geese eat the underground stems. Cattails furnish excellent cover and nesting sites for various ducks, Long-billed Marsh Wren, Red-winged Blackbird, and Green-winged Teal.

Today the cattail is popular with foragers who know the many fine products it produces. The rhizomes can be peeled of their spongy outer layer, dried, ground, and sifted for flour. To my experience, the best flour is made without drying the stem first. Peel the rhizomes to their starchy core,

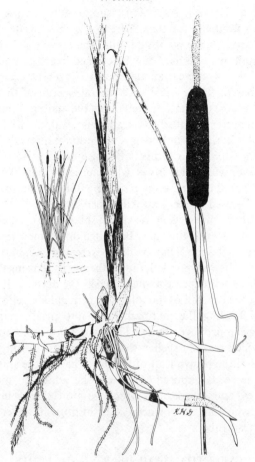

Fig. 46. Common CATTAIL *(Typha latifolia)*.

crush them in water (or process in a blender), strain out the fibers, and let the flour settle to the bottom of the water. Decant the water and use the flour either wet or dry. Cattail flour may be combined with other types of flour.

The little white shoots that come off the underground stems will be next year's cattails. These pointed, curved shoots are tender and delicious raw or cooked like any vegetable. The young cattail plants about 12-in high are known as Cossack Asparagus, supposedly because of the Russian fondness for them. Grasp the stalk inside the two outer leaves of the plant and pull. The lower portion of the inner stalk may be used raw or cooked like asparagus. The base of the mature cattail also yields a vegetable resembling cucumber in flavor—just slice the inner core of the larger plants. Where the mature stem meets the underground stems (often incorrectly called roots) there is a walnut-sized ball of starchy material resembling a small potato; this is the cattail heart—a delicacy to foragers because of its versatility and mild flavor which blends well with sauces or meats. The heart may be pared out and used in making flour or potato dishes. It may be used in recipes for Bulrush also.

In early summer when the cattail spikes are still green and wrapped in leaves, they may be boiled or steamed and eaten like corn on the cob. They are dry, best served with butter, but their flavor is very good. Scraping them off and using them in a casserole with cheese and egg masks the dryness. If left to mature, the top half of the spike will turn golden yellow with pollen. Bending the matured heads into plastic bags and shaking can produce about 2 Tbsp of delicate and nutritious flour which makes excellent baked products. Burning off a portion of the brown fluff of the lower half of the spike will leave the seeds exposed. They can be added to muffins, pancakes, and breads. For Mock Potato Salad, follow the recipe given under Arrowhead, substituting cattail hearts for Arrowhead tubers.

CREAMED CATTAIL HEARTS AND SHOOTS

6 cups cattail hearts, and shoots, diced	1 1/4 tsp salt
	1/4 tsp pepper
2 Tbsp butter or margarine	2 cups milk
	lemon juice
2 Tbsp flour	

Steam or boil cattail hearts and shoots until easily pierced with a fork (about 25 minutes). Melt butter or margarine, stir in flour, salt, and pepper, and cook, stirring constantly, for 1 minute. Add milk and cook until thickened. Add a squeeze of lemon and cattails and simmer for 5 minutes. Serves 6.

COSSACK ASPARAGUS WITH CHEESE SAUCE

6 cups young
  cattail plants
3 Tbsp butter
  or margarine
3 Tbsp flour

2 cups milk
1/2 tsp salt
dash of pepper
1/2 cup shredded
  cheddar cheese

Steam or boil cattails for 10 minutes. Melt butter or margarine, stir in flour and milk gradually, then add cheese until smooth. Salt and pepper to taste. Pour over cattails. A simpler sauce may be made by adding the shredded cheese directly to the hot, drained vegetable, covering, and letting stand until cheese is melted, then stirring. Serves 6.

GREEN CATTAIL SPIKE CASSEROLE

2 cups green cattail
  spikes, scraped
3 eggs, slightly beaten
1 1/2 cups shredded
  cheddar cheese
3 cups milk
6 Tbsp chopped onion
  and pimento (optional)

1/2 tsp salt
dash of pepper
1/3 cup bread crumbs
  (or cracker crumbs)
2 Tbsp butter
  or margarine

Preheat oven to 350° (mod.). Blend first 7 ingredients in buttered baking dish. Sprinkle with bread crumbs and dot with butter or margarine. Bake for 45 minutes. Serves 6.

SUNSHINE PANCAKES

3/4 cup sifted
  cattail pollen
3/4 cup flour
1 tsp baking powder
1/2 tsp salt

1 Tbsp brown sugar
  or honey
2 eggs
1 Tbsp melted butter
  or oil
1 cup milk

Mix all ingredients until smooth. For fluffier pancakes, separate the eggs and beat the egg whites, folding in last. Cook on a hot ungreased griddle. Serves 4.

74. CELERY
Apium graveolens (fig. 47) Family: Umbelliferae

Description: The familiar plant of gardens and super-markets. Perennial herb 5-12 dm tall with divided basal leaves. Tiny flowers are in umbrellalike inflorescence and white in color. Smell is unmistakably like the garden variety, but the taste is much stronger. The ribbing on the stalk is also distinctive. Blooms May through July.

Distribution and Habitat: Wild Celery is common in wet places at low elevations.

Uses—Past and Present: Celery was introduced from Europe where it has a long history as a medicinal herb. Indians used this plant as a potherb. The Cahuillas made a tea of Celery for kidney complaints. The tops of the plants can absorb dangerous levels of nitrates and should be discarded. The wild plant, though easily recognized by its smell, has a much stronger aftertaste than the cultivated plant. For this reason it is best mixed with other strong flavors or cooked rather than eaten alone or raw. Ancient Greeks used Celery as an award at sports events. Abyssinians stuffed pillows with Celery leaves as a cure for headache. Positive identification of the plant is imperative, since some other members of the Umbelliferae are toxic to humans (see pl. 8e).

SWEET-SOUR CELERY

3 cups sliced	1/2 tsp salt
Celery stalks	dash of pepper
1 egg, beaten	2 Tbsp vinegar
2 Tbsp flour	1 cup water
2 Tbsp sugar	1/4 cup sour cream

Boil or steam celery and drain. Blend egg, flour, sugar, salt, and pepper in saucepan and add vinegar and water. Bring to a boil, remove from heat, and add sour cream. Mix with celery. Serves 6.

CREAM OF CELERY SOUP

2 1/2 cups milk
1 Tbsp flour
2 Tbsp butter
  or margarine
1 tsp salt

dash of pepper
1 cup coarsely
  chopped Celery
1/3 cup diced onion
1 Tbsp lemon juice

Place all ingredients in a blender and blend until smooth.
Pour into a saucepan and simmer for 5 minutes. If blender is
not available, simmer all ingredients except sour cream, until
tender. Stir in sour cream before serving. Serves 4.

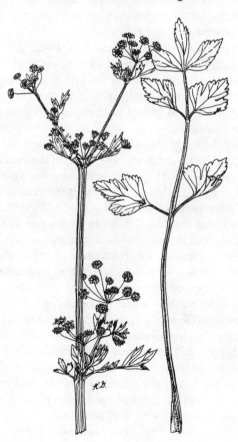

Fig. 47. CELERY *(Apium graveolens).* 1/2 X.

75. Sweet FENNEL
   (also Anise, Finocchi) *Foeniculum vulgare* (pl. 5a)
   Family: Umbelliferae

Description: Erect perennial herb with anise odor. Leaves are dissected into linear divisions, loosely resembling feathers. Plant is 1-2 m high with a white powder coating. Flowers are yellow and occur in umbels. The odor of licorice is perhaps the best overall characteristic by which to identify this plant. Blooms May through September.

Distribution and Habitat: Common in waste places especially in southern and central California. Particularly abundant in coastal regions, occurring frequently on cliffs and sand dunes opposite beaches.

Uses—Past and Present: Originally from the Mediterranean area, Fennel now occurs all over the world. Ancient Egyptians, Hindus, and Chinese used its seeds for spices. The Romans—then as now—cultivated it for its seeds and edible shoots. The fragrant fronds were made into garlands with which to crown victorious warriors. It is used extensively today in French and Italian cooking, and there are selected horticultural forms available. The leaves of the Fennel can be used in sauces or cooked as a vegetable. The young stems before they flower, when still tender, are cooked like Celery, much of the anise flavor being dissipated in the cooking. Raw stalks are added to salads. The seeds are dried or gathered dry and used in fruit dishes, breads, and cookies. Care must be taken when gathering seeds from dried plants to distinguish them from Poison Hemlock, which they resemble in this state. The seeds should have a definite licorice odor. One can easily see dried feathery leaves on Fennel, whereas hemlock leaves resemble the lacy leaves of a carrot (see pl. 8e). Tea can be made by steeping Fennel leaves in hot water.

It is said the mission fathers of California used to spread Fennel stems on the floor of the missions so when they were bruised by the feet of the congregation, the pleasant aroma would fill the rooms. Anise is a good bird plant of southern California. It has flowers and seeds and attracts aphids, therefore providing varied foods for many birds.

FENNEL IN LEMON-BUTTER SAUCE

2 cups tender   1 Tbsp lemon juice
  Fennel stems   3 Tbsp butter or margarine

Gather only the flexible stems at the tips of the plant or those just emerging from the ground. Larger stems may be peeled of their tough rind and the core only used. Steam or boil the stems until easily pierced with a fork. Drain while still hot, add butter and lemon juice, stir vigorously, cover, and let steep for 2 minutes before serving. Much of the licorice taste will dissipate in cooking, so this dish is not strong. Serves 4.

FENNEL AND FRUIT SALAD

1 heaping cup tender   1/3 cup chopped walnuts
  raw Fennel stems, sliced   1 cup vanilla yoghurt,
2 red apples, chopped   thinned with 1 Tbsp
1/2 cup raisins,   lemon juice
  softened in hot water
  for 5 minutes

Drop Fennel, fruit, and nuts into yoghurt as you prepare them to keep them from turning brown. Stir. Chill. Serves 6.

FENNEL COOKIES

1 Tbsp Fennel seeds,   1/2 tsp baking soda
  dried (if seeds are   1 egg
  chewy, chop them)   2 cups flour
2 Tbsp boiling water   3/4 cup butter
2/3 cup brown sugar   or margarine

Preheat oven to 400° (mod. hot). Combine seeds and water and let stand while mixing dough. Cream butter with sugar and baking soda. Beat in egg. Drain seeds and add with flour. Mix well. Shape dough into 1/2-in balls and place on ungreased baking sheet. Bake for 10 minutes. Makes about 4 dozen. For a prettier cookie, roll in powdered sugar after removing from the oven. Dough may be made ahead and refrigerated prior to baking. These are delicate and delicious.

76. ICEPLANT

Gasoul crystallinum formerly *Mesembryanthe-mum crystallinum* (fig. 48) Family: Aizoaceae

Description: A low succulent annual with prostrate branches 2-6 dm long. The branches and leaves (which soon drop off) have numerous large watery bubbles on their surfaces; leaves, when present, ovate, 2-10 cm long. Has 5 sepals, unequal in length; many petals, linear, white to reddish, 6-8 mm long. Flowers March through October.

Distribution and Habitat: Saline places such as salt marshes and seashore dunes, Coastal Strand, Coastal Sage Scrub, along the coast from Monterey Co. to Baja California. Another species, *Carpobrotus edule* (formerly *Mesembryan-themum edule*), which carries the common name Hottentot Fig or Sea Fig, is also edible and tastes much the same, though not as mild. It is distinguished by its typical triangular leaves on the cultivated plants and is often mistakenly called Iceplant.

Uses—Past and Present: Both species mentioned above were apparently introduced from Africa and thus not used by the early Indians. Iceplant was advertised in American seed

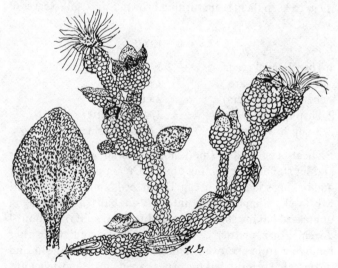

Fig. 48. ICEPLANT *(Gasoul crystallinum).* 3/4 X.

lists of 1881 as a desirable vegetable for boiling like spinach or for a garnish. In its native land, the entire plant of a similar species is beaten, fermented, and chewed. The leaves can be used as a substitute for cucumber in making pickles, but I have found them best when added fresh to salads. Tossed alone or with other greens in the following dressing, Iceplant makes a gourmet dish.

DRESSED ICEPLANT

1/4 lb bacon, diced, fried, and drained	1/4 cup buttermilk
2 eggs, beaten	1/2 cup water
2 Tbsp brown sugar	dash of pepper
1 tsp flour	1 qt Iceplant leaves

Combine all ingredients except Iceplant and cook, stirring constantly, until dressing thickens. Pour over Iceplant and toss. Serve while warm. Serves 4.

77. Feather Boa KELP.
Egregia laevigata (pl. 5b) Division: Phaeophyta

Description: Many-branched (6-25) perennial kelp, each branch a narrow stemlike growth. Outer edges of all branches are densely fringed with blades of various shapes and sizes, about 0.5-1.5 dm long. Some short blades develop into spindle-shaped air bladders. Surface of both stipe and blades smooth or covered with small tubercles. Mature plants 5 m or more in length, with branches 1-1.5 m long. Color is brown, turning green when cooked.

Distribution and Habitat: Growing on rocks between 0.5 and -0.5 ft (15 and -15 cm) tide levels from central California to Baja California. A similar species found further north is *E. menziesii*.

Uses—Past and Present: Like other large brown algae, this species is used for fertilizer. For many centuries seaweeds have been used to increase crop yields in countries with extensive seacoasts. The fertilizing properties of seaweeds are about the same as those of barnyard manure; tests have shown that seaweeds contain a greater proportion of potassium salts, about the same proportion of nitrogen, but a smaller amount of phosphoric acid.

A few seaweeds appear in the dried state in our markets, but their uses are almost unknown to the general public. In parts of Europe, seaweed is more highly appreciated than here, but it is in the Orient, in China and Japan, that it is prized for the good food that it is. A search of the literature indicates that though most people are unaware of the delicious and healthful seaweeds, there does exist an adventuresome group of people who have discovered that our entire seacoast is lined with many edible marine algae. To prepare for eating, kelp may be rinsed and eaten raw cooked, or dried for future use. I find both the stems and blades quite palatable and even delicious when pickled. Note that there are no poisonous seaweeds known.

PICKLED KELP

1 cup kelp (or other seaweed) stipes and blades, cut in 6- to 8-in pieces	1/4 cup diced onion
	3/4 cup sugar
	1/2 lemon, sliced
	1 cinnamon stick
1/2 cup vinegar	4 whole cloves
1/4 cup water	pinch of mace

Place kelp in a kettle and bring to a boil for about 5 minutes. (Enjoy their green color now!) Drain, rinse, and slice into bite-sized pieces. Young kelp will be tender at this stage. If not tender, they may require boiling for up to 1 hour. Add rest of ingredients and cook, covered, for 1/2 hour. Seal in jars or refrigerate. Serves 2.

RAW VINEGARED KELP (SUNOMONO)

1/4 cup vinegar	1/4 tsp salt
1 tsp soy sauce	1 cup kelp pieces
1 Tbsp sugar	

Combine vinegar, soy sauce, sugar, and salt. Wash kelp, pour boiling water over it, and let stand. Drain and serve with sauce. Kelp may also be marinated in the sauce. Serves 2.

78. GIANT KELP
Macrocystis pyrifera (fig. 49) Division: Phaeophyta

Description: Many-branched, common kelp in which each
blade is attached to the main stipe by an air bladder. A single
blade is up to 8 dm long and 4 dm broad. The terminal blade
of each branch is broad with 10-20 young blades in
progressive stages of differentiation by an asymmetrical
splitting from base to apex. The mature lateral blades have
toothed margins and an irregular, corrugated shiny brown
surface. A single plant may be 25 m long.

Distribution and Habitat: Growing in extensive stands on
rocks 2-3 ft (6-9 m) below the surface of the water. Upper ends
of branches grow to surface of water. From Alaska to Baja
California. A smaller species is *M. integrifolia,* found in the
waters off central and northern California.

Uses—Past and Present: In California this perennial kelp
is being used extensively to make algin, which is widely used
in the manufacture of ice cream, chocolate milk, and jellied
puddings to help keep them creamy and smooth. It is also
used in pharmaceutical products such as hand lotions. The
recipes for Feather Boa Kelp are delicious when made with
Giant Kelp. Young or tenderized blades may be rolled
individually around vinegared rice, as in the Sushi recipe
given for Kombu.

79. KOMBU
(also Wrack, Kelp) *Laminaria farlowii* (fig. 50)
Division: Phaeophyta

Description: Unbranched stipe is cylindrical in cross-
section, dark brown in color. Stipe is fairly short, up to 6 cm
long and 1.5 cm in diameter. Blade is up to 3 dm long, linear
with irregularly wrinkled surface.

Distribution and Habitat: Growing on rocks between 0.5
and -0.5 ft (15 and -15 cm) tide levels. It ranges from central to
southern California. Other species of *Laminaria* range the
entire coast. One species, *L. saccharina,* is sweet when
chewed.

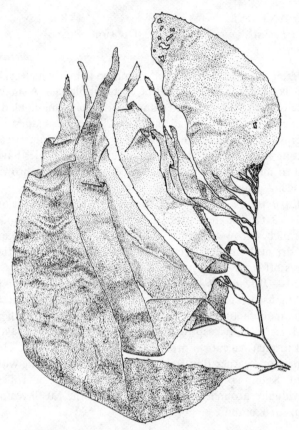

Fig. 49. GIANT KELP *(Macrocystis pyrifera).* 1/2 X.

Uses—Past and Present: The large brown algae, to which this species belongs, were among the sources of commerical iodine during World War I. It is a source of algin (discussed under Giant Kelp). The blades or stipes may be rinsed and used fresh or dried in the sun and preserved for later use. Other recipes include those for Feather Boa Kelp.

KOMBU SOUP

1 Tbsp oil   1/2 cup green beans,
1 small onion, sliced   sliced

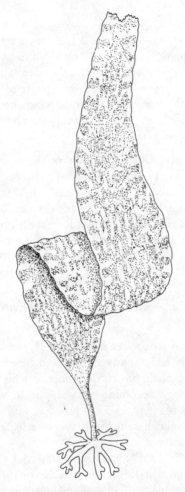

Fig. 50. KOMBU *(Laminaria farlowii)*.

1 Tbsp soy sauce
2 cups beef broth
1/4-in slice fresh ginger
  minced

1 cup water
3 strips (2x5 in) dried
  Kombu blades,
  crumbled

Heat oil in saucepan over medium heat, cook onion and green beans until onion is translucent. Stir in remaining ingredients and simmer for 5 minutes. Serves 4.

SUSHI

1/4 cup vinegar	2 1/2 cups water
3 1/2 Tbsp sugar	2 cups white rice, washed in
2 1/2 tsp salt	cold water and drained
1 1/2 Tbsp sake or	2-in cube dried Kombu
1 Tbsp pale dry sherry	blades, washed under
1/2 tsp monosodium	cold water
glutamate (MSG) (optional)	

Combine vinegar, sugar, salt, sake or sherry, and MSG in a saucepan. Bring to a boil then cool to room temperature. Combine water and rice and let soak for 30 minutes. Add Kombu and bring to a boil. Cover and simmer for 10 minutes or until rice has absorbed all water. Remove from heat and let stand for 10 minutes. The seaweed may be removed at this point. Combine dressing with rice, and serve cold. Serves 4.

80. LAVER

(also Nori) *Porphyra perforata* (fig. 51) Division: Rhodophyta

Description: A lettucelike blade of steel gray to brownish-purple sets this leafy seaweed off from Sea-Lettuce. The single, flattened blade has ruffled margins which are bright red when young, becoming grayish-purple when older. Frequently, the blade is 4.5-5 dm broad as well as long. The blade is very fragile, having but 1 layer of cells, and will thus very often be tattered.

Distribution and Habitat: Common on rocks between 3.5 and 2.0 ft (1.7 and .6 m) tide levels as well as on the other algae from Alaska to southern California. Numerous other species of *Porphyra* exist and may be used similarly.

Uses—Past and Present: Laver is cultivated by the Japanese who call it Nori. Chinese-Americans process many pounds of it each year, exporting some to China. In Puget Sound, the Indians still use *Porphyra* as flavoring for soups and meats. Gourmet recipes abound. Other seaweed recipes that are compatible with this plant are those for Feather Boa Kelp, Kombu, Sea Lettuce, and Sea Whip.

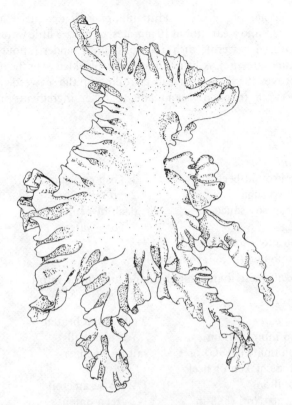

Fig. 51. LAVER *(Porphyra perforata)*.

STUFFED LAVER

1 lb sausage or
  ground beef
1/4 lb mushrooms,
  chopped
1 medium onion, chopped

2 cups cooked rice
1 Tbsp plus 1 tsp
  soy sauce
dried Laver (or other
  seaweed)

Fry sausage or ground beef, drain (reserving oil), and set aside. Return oil to pan and sauté mushrooms and onion. Combine rice, sausage or ground beef, mushrooms, onion, and soy sauce. Drop seaweed into boiling water for 1 minute to make soft and pliable. On each piece about the size of a folded napkin, place 1 Tbsp filling. Roll up, place in a steamer, and steam about 10 minutes, or place a little water in a covered casserole dish and roast until tender (about 20 minutes at 375°). A little lemon juice sprinkled over the dish improves the flavor. Instead of rolling up the seaweed, you may wish to chop it and add to other ingredients in a casserole.

NORI SOUP

Meatballs:

1 lb ground beef	1 Tbsp cornstarch
1/2 tsp sugar	2 tsp salt
2 Tbsp soy sauce	dash of pepper
1 egg	2 Tbsp water
1 tsp monosodium glutamate (MSG) (optional)	

Mix all these ingredients by hand and shape into balls.

Soup:

1 qt water	2 eggs, well beaten (or meatballs)
1 tsp monosodium glutamate (MSG) or 1 cube chicken or beef bouillon	salt and pepper to taste
	1/2 tsp sesame oil
2 sheets Nori (8x8-in squares of double or triple thickness)	1-2 green onions, chopped

Boil water with MSG or bouillon cube. Add Nori and stir with a fork to loosen. When Nori is soft, add eggs or meatballs. Boil for 10 minutes or until meat is done, and add salt and pepper to taste. Sprinkle with sesame oil and green onions, and serve hot. Serves 4.

81. Stinging NETTLE
   (also Nettle) *Urtica dioica* ssp. *holosericea* (pl. 5c)
   Family: Urticaceae

Description: Perennial herb with stout, unbranched stems bearing bristly stinging hairs. Stems 1-2.5 m tall, lanceolate leaves 5-12 cm tall and borne opposite each other on the stem, softly pubescent, grayish beneath. Flowers hang loosely in clusters at base of leaves, are tiny and green. Blooms in summer.

Distribution and Habitat: This species is common in damp places below 9000 ft (2800 m) throughout the state. Four other species are also found in California and all are edible. *U. urens,* a common orchard weed, is particularly tasteful.

Uses—Past and Present: Nettle leaves were used by the Indians for a raw vegetable or boiled as greens. Stem fibers were used to make bowstrings and in basketmaking. Nettles were applied to various aching areas, such as rheumatic legs and arms. The boiled roots were used in early times in Europe to obtain a yellow dye. To foragers, nettles are not only delicious but healthful, containing significant amounts of vitamins A and C and protein. The rash that follows an encounter with the stinging hairs usually lasts only a short time, but they should be gathered with care, using gloves or scissors. The sting contains formic acid and is similar to being stung by certain ants. Cooking removes this unpleasant aspect. Young nettles are best steamed or boiled, but I find the leaves of the larger plants are the only edible part. Nettles can be substituted in any recipe calling for spinach or chard. Partly due to its high nitrogen content, the nettle is an active decomposer and humus maker and is often used as a catalyst to ferment compost heaps by commercial houses that sell purely organic fertilizer. Nettle plants are reportedly used as a food source by the Brown Towhee.

CAMPFIRE NETTLES

Lightly cook tender nettle parts in a little boiling water for a minute or two, until they are tender. Drain, add butter and lemon, if desired. Nettles are a delicious, if unusual, substitute for spinach.

SPECIAL NETTLES

1 lb nettle leaves and young stems	1 tsp salt
	6 slices bacon, diced

1 cup diced cooked beets	dash of pepper
2 Tbsp vinegar	2 hard-cooked eggs,
1/2 tsp salt	diced

Sprinkle nettles with 1 tsp salt and cook in very little water. Drain and chop coarsely. Fry bacon in skillet; add beets and heat a few minutes. Add vinegar, 1/2 tsp salt, pepper, nettles, and eggs. Heat a few more minutes and serve. Serves 5.

Steamed or boiled nettles may be added to cream of mushroom soup or cooked separately and substituted for onion in the recipe for Cream of Celery Soup given in this section.

82. NUT-GRASS

(also Chufa) *Cyperus esculentus* (fig. 52) Family: Cyperaceae

Description: Perennial herb with unbranched, triangular stem which is leafy at base like grasses. Underground stems are scaly and bear tubers. Plant is 1.5-5 dm tall with the leaves about as long and 3-10 mm wide. Has 2-6 leaves, which are found under the umbellate flowering head. Small grasslike flowers are straw-colored. Blooms June through October.

Distribution and Habitat: This species is a common weed of fields and wet places at low elevations throughout the state. Other species do not always bear tubers, but, similarly, the young plants and leaf bases are edible.

Uses—Past and Present: Nut-Grass must have been prized in ancient times in other countries, for its tubers have been found in Egyptian tombs as old as 2400 B.C. In California, Paiute Indians pounded the tubers of the plant together with tobacco leaves and applied this to treat athlete's foot. The edible tubers and Nut-Grass seeds are utilized by numerous waterfowl, small mammals, and songbirds, such as Coot; Mallard; Pintail, Blue-winged, Cinnamon, and Green-winged Teal; crow; Horned Lark; and kangaroo rat. Nut-Grass is common in mud flats that are covered by water during the cold seasons of the year. In these locations, the tubers can be easily retrieved by ducks when other food sources are not so readily available. Nut-Grass tubers can be

Fig. 52. NUT-GRASS *(Cyperus esculentus).*

soaked for 2 days in cold water, cleaned, and pulverized with water and sugar to yield a sweet milky drink much relished in Spain. Peeled and roasted tubers are gound up to use as a coffee substitute or as flour. The flour is sweet and makes delicious cookies and other baked goods. To roast tubers for flour, place in a slow oven until they are brittle, rather than spongy. The tubers may be used raw in salads or steamed and eaten with butter and seasonings. The base of the plant stem makes a nice raw vegetable, also.

CHUFA COOKIES

1/2 cup butter or margarine	1 cup white or wheat flour
1 1/3 cups brown sugar	1/2 tsp salt
2 eggs	1 tsp baking powder
3/4 cup Nut-Grass flour	1 cup chopped almonds (or other nuts)

Preheat oven to 350° (mod.). Cream margarine with about 1/4 cup brown sugar. Beat remaining sugar with eggs. Add these together and blend. Mix in dry ingredients and stir in nuts. Drop onto greased cookie sheet and bake for 8-10 minutes. Makes 2-3 dozen.

SUGAR-BROWNED CHUFA

3 doz cleaned Nut-Grass tubers	6 Tbsp sugar
1/4 cup butter or margarine	2 tsp salt

Boil or steam tubers until tender (about 15 minutes). Drain. Melt butter or margarine in skillet, stir in sugar, and cook until light caramel. Add tubers and stir. Sprinkle with salt. Serves 6-8.

83. Parish's PICKLEWEED

(also Glasswort, Samphire, Chicken Claws) *Salicornia subterminalis* (fig. 53) Family: Chenopodiaceae

Description: Perennial herb with widely spreading, somewhat fleshy, stems, 1.5-3 dm high. Leaves are reduced to tiny scales. Stems are jointed with inconspicuous flowers from August to November. Older plants often turn reddish.

Distribution and Habitat: Forms mats in salt marshes and low alkaline places such as Coastal Salt Marsh, Alkali Sink, Coastal Sage Scrub. All four species occur throughout the state and may be used similarly, though this particular species tastes best to me.

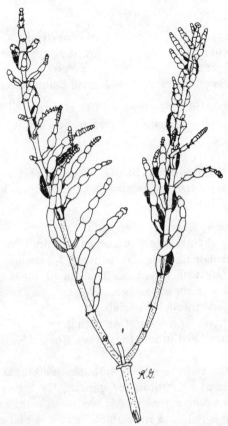

Fig. 53. Parish's PICKLEWEED *(Salicornis subterminalis).* 1/2 X.

Uses—Past and Present: Pickleweed seeds, which are available in summer, were a favorite with the Cahuillas. The Indians of Utah and Nevada gathered the seeds and ground them into flour. Pickles were made by colonists by first boiling them in salted water then soaking them in vinegar. The name glasswort comes from their high soda content, which was of value to the glass and soap manufacturers. The bright green, tender tips of the plant are used in seaside salads and pickles. It can be used as a potherb, but is not as tasty in that use.

GLASSWORT PICKLES

1 qt white or wine vinegar	3 bay leaves (if California Bay, use only 2 leaves)
1/2 cup sugar	
2 Tbsp pickling spices	1 small onion, sliced

Boil all ingredients together for 10-15 minutes and pour hot over fresh glasswort stems in glass jars. Seal and let stand for 1 month.

84. Halberd-leaved SALTBUSH
    (also Orach) *Atriplex patula* ssp. hastata (fig. 54)
    Family: Chenopodiaceae

Description: Annual, outer surface of stems and leaves glabrous or slightly mealy, branches 3-10 dm long. Leaves green, lanceolate to triangular, petioled, 3-8 cm long. Lower leaves dentate with broad basal angles or lobes that are rounded or straight at the base. Flowers small and green, produced June through November.

Distribution and Habitat: Common in moist saline places along the coast and in interior valleys, Coastal Salt Marsh, Alkali Sink.

Uses—Past and Present: As with most saltbushes, Orach may be used as a substitute for spinach. The leaves have a natural salt content that blends well with a little lemon. Although this Saltbush is palatable raw, it is best cooked, without salt, as a potherb. It may be used in recipes for Lamb's Quarters, amaranth, and New Zealand Spinach. Some species of *Atriplex* are listed on the California Native Plant Society's rare and endangered plant list and should be used only in extreme emergencies. Saltbush is utilized in many ways by animals and birds. The seeds are eaten by many birds, including wild geese, which also eat the leaves and stems. Seeds and leaves are eaten by Antelope Ground Squirrel, pocket mouse, and kangaroo rat. Twigs and foliage are eaten by various hoofed browsers. See General Index for other saltbush species found elsewhere.

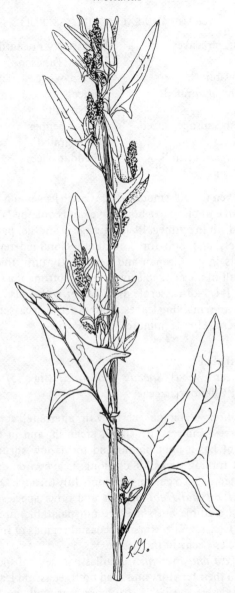

Fig. 54. Halberd-leaved SALTBUSH *(Atriplex patula* ssp. *hastata).* 3/4 X.

SAUSAGE AND SALTBUSH CASSEROLE

6 cups Orach leaves,
washed
1/2 cup water
1/4 tsp freshly grated
nutmeg
1 lb Polish sausage, sliced
1/4 cup beer
1/4 cup cream or milk
2 Tbsp flour

1/4 tsp dry mustard
3 drops Tabasco
1/4 tsp Worcestershire
sauce
1/4 tsp salt
dash of pepper
2 cups grated
cheddar cheese

Preheat oven to 350° (mod.). Place Orach leaves and water in a saucepan and bring to a boil. Cover and cook for 7 minutes. Drain and stir in nutmeg. Place sausage in another pan, cover with water, and boil for 3 minutes. Combine remaining ingredients in a saucepan and heat gently until thickened. Spread half of the cooked leaves on the bottom of a shallow, buttered, 1 1/2-qt casserole dish. Cover with sausage slices and top with remaining leaves. Pour sauce over casserole and bake, covered, for 15 minutes. Serves 4.

85. SEA BLITE
(also Seep Weed) *Suaeda californica* (fig. 55)
Family: Chenopodiaceae

Description: Perennial herb with alternate, somewhat fleshy, leaves. Flowers are small, greenish, and in the leaf axils. Plant has a light bloom and forms low shrubs with somewhat trailing stems 3-8 dm long. Leaves are numerous and crowded, 1.5-3.5 cm long. Blooms July through October.

Distribution and Habitat: This and other species of Sea Blite (all are edible but some are not palatable) are found mainly in Coastal Salt Marsh in coastal regions of the state. Other species occur in the desert.

Uses—Past and Present: Cahuillas had several species of Sea Blite in their territory and used both seeds and leaves for food. One desert species, *S. fruticosa,* was made into a hair dye by boiling the leaves and mixing them with clay. Baskets were also dyed with *Suaeda* coloring. Some species are very

Fig. 55. SEA BLITE *(Suaeda californica)*. Actual size.

salty and require rinsing in a change of water or boiling for 5 minutes to make them palatable. Sautéed or steamed Sea Blite loses its gray-green color and turns bright green. The upper, tender stems, leaves, and flowers are all pleasant when cooked. Served with butter, bacon, or a vinegar dressing, Sea

Blite is a tasty vegetable. It may also be mixed with cream of celery or cream of mushroom soup. It makes a good salad if cooked and mixed with dressing and hard boiled eggs.

SEA BLITE AND SOUR CREAM SALAD

4 cups Sea Blite leaves, stems, or flowers, cooked, drained, and chopped
1 Tbsp sugar
1/2 tsp salt

1 cup sour cream
3 Tbsp grated onion
2 Tbsp vinegar or lemon juice
2 hard-cooked eggs, diced

Blend all ingredients together and chill for 2 hours before serving. Serves 4-6.

GARNISHED SUAEDA

4 cups chopped Sea Blite
6 slices bacon, chopped and fried crisp

2 Tbsp Parmesan cheese
2 Tbsp wine vinegar

Boil or steam Sea Blite. Drain and combine with remaining ingredients. Add a little bacon fat, if desired. Serve hot. Serves 4.

86. SEA LETTUCE
(also Green Laver) *Ulva lactuca* (fig. 56) Division: Chlorophyta

Description: A bright green sheet or ribbonlike alga with wavy margins. It has no stem.

Distribution and Habitat: Found on most seacoasts and clinging to docks and piers. Some plants may be found in backwater areas, due perhaps to tidal action.

Uses—Past and Present: This large member of the green seaweeds has been used for food in England, Ireland, and Scotland for centuries. It is tough, though tasty, raw and must be chopped into small pieces to be added to salads. Upon drying, however, it loses this elasticity and may be crumbled and used as a garnish for salads or added to cooked

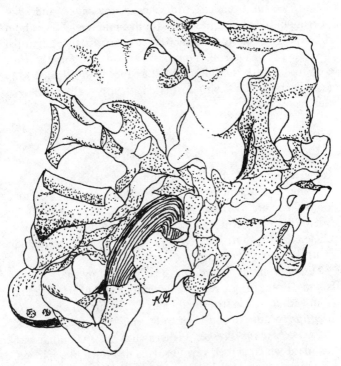

Fig. 56. SEA LETTUCE *(Ulva lactuca)*. Actual size.

dishes such as rice and fish. Although it is not as desirable, it may be used much as Laver. Various species of Sea Lettuce are reportedly eaten by Canada Geese and are one of the principal foods for brant. Sea Lettuce is also eaten by Coot, Mallard, Pintail, Common and Surf Scoter, Shoveler, Avocet, and Sora Rail.

INSTANT SUSHI

2 cups rice cooked
  in onion soup (substitute
  1 can soup for
  1 cup water)
10 sheets Sea Lettuce
  dried and resoaked for
  a few minutes

2 Tbsp wine vinegar
1/2 tsp soy sauce
2 eggs, beaten

Cook rice until tender. Add vinegar, soy sauce, and eggs, stirring until the eggs have set (about 5 minutes). Place about 1 Tbsp of this filling onto each piece of Sea Lettuce and roll it up like a cabbage roll. You may, instead, chop up the *Ulva* and toss it with the filling. A little sauce from Pickled Kelp (see Feather Boa Kelp) sprinkled on top really adds zest. Serves 4.

87. SEA ROCKET
Cakile maritima (pl. 5d) Family: Cruciferae

Description: Fleshy annual with leaves deeply lobed, 4-8 cm long. Sepals 3 mm long; petals pink to purplish, 8-10 mm long. There are 4 sepals, 4 petals, and 6 stamens, as in all the Mustards. Fruit is a slender, podlike structure known as a silicle, 12-15 mm long, with two divergent triangular protruberances near its tip.

Distribution and Habitat: Beach sand on the Channel Islands, Point Mugu, near El Segundo, Los Angeles Co., and from Monterey Co. to Mendocino Co. A similar species, *C. edentula,* occurs in southern California.

Uses—Past and Present: Fishermen are often familiar with this plant which is used raw in salads or boiled like spinach. Its peppery horseradish taste is best toned down with other greens or ingredients, however, rather than eaten alone. Young plants or leaves are best, though older leaves may be used if boiled longer. Buds, immature pods, stems, and leaves are used. They make a nice accompaniment to meat sandwiches.

88. SEA WHIP
(also Bull Kelp, Ribbon Kelp) *Nereocystis leutkeana* (pl. 5e) Division: Phaeophyta

Description: Perhaps one of the most conspicuous seaweeds in California waters, the Sea Whip may have a stipe 10-15 m long. Blades are held at the surface by air bladders. When tide is slack, ribbonlike blades hang down, leaving only air bladders on the surface. Grows in large fields or beds far out from the shore, occasionally closer to shore. Lower end of stipe is solid; upper end hollow. Air bladder at tip of stipe may be 8-15 cm long. From this bladder, blades emerge 3-4 m long and 3-4 cm wide. Blades of mature plants are shiny and leathery, while younger plants have thinner, shinier brown blades.

Distribution and Habitat: Southern California to Alaska below the water surface, although younger plants may be torn loose and found ashore.

Uses—Past and Present: Ribbon Kelp is reportedly used in the preparation of pharmaceutical supplies, dairy products, poultry feeds, and glazing and finishing agents. It is also an excellent source of potash salts. Since all seaweeds are edible, recipes for others found in this text may be interchanged rather freely. See Feather Boa Kelp, Giant Kelp, Kombu, Laver, and Sea Lettuce.

EBI ISOBE YAKI (SHRIMP WRAPPED IN SEAWEED)

2 tsp sake or
  dry sherry
2 tsp soy sauce
monosodium glutamate
  (MSG) (optional)

1 doz medium-sized un-
  cooked shrimp, shelled,
  deveined, and cut to
  straighten them out
1 doz 6-in blades Sea Whip,
  dried and soaked for a
  few minutes

Combine sherry, soy sauce, and MSG. Marinate shrimp in sauce for several hours or overnight. Roll shrimp tightly in Sea Whip blades and arrange on a broiling pan. Broil 3-in from heat for about 3 minutes or until shrimp is firm. Cut into 1/2-in bits and serve hot with more sauce if desired. Serves 3-4.

89. New Zealand SPINACH
  Tetragonia tetragonioides formerly *T. expansa*
  (fig. 57) Family: Aizoaceae

Description: Succulent annual with many spreading branches. Leaves and stems have small crystalline bubbles.

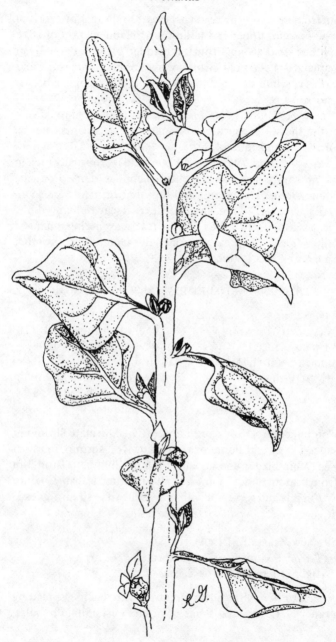

Fig. 57. New Zealand SPINACH *(Tetragonia tetragonioides)*. Actual size.

Alternate leaves are triangular-ovate, contracted at the base to a winged petiole. Flowers yellow-green, 1.5-2.5 mm long, solitary in angles of leaves. Fruit 8-10 mm long with 2-5 horns. Blooms April through September with fruit following.

Distribution and Habitat: Naturalized along beaches and near salt marshes of the coast over the entire state.

Uses—Past and Present: New Zealand Spinach grows in Japan, Australia, and New Zealand as well as various other locations around the world. It is cultivated in South America as a potherb and has escaped in California. This plant was first discovered by the sailors of Captain Cook's expedition around the world. It was cultivated in England in 1821 and in France in 1824. It is used as a potherb in Tongatabu but not in New Zealand. Foragers enjoy its mild flavor and crunchy texture in salads and as a potherb. It has no strong briny taste and can be served with a variety of sauces and dressings. The plant is a favorite of mine just lightly steamed with lemon and butter or a slightly tart sauce such as Hollandaise.

NEW ZEALAND SPINACH WITH HOLLANDAISE SAUCE

4 egg yolks	6 cups spinach
1/2 cup butter	leaves and stems, lightly
3 tsp lemon juice	steamed or boiled
	salt and white pepper
	to taste

Place egg yolks in top of double boiler and beat slightly. Add about 1/3 of the butter and place over hot, but not boiling, water. Cook, stirring rapidly, until butter melts. Repeat until all butter has been added. Remove top of double boiler from heat and stir mixture rapidly about 2 more minutes. Stir in lemon juice, 1 tsp at a time, and add salt and pepper. Place back over hot water and stir 2-3 minutes. (If the sauce curdles, immediately beat in 1-2 Tbsp boiling water.) Serve over Spinach. Serves 6.

PLANTS TREATED ELSEWHERE WHICH MAY ALSO
OCCUR IN WETLANDS
(Refer to General Index for page numbers)

Wild Grape

Wild Mint

Onion

Peppermint

Salmonberry

Sheep Sorrel

Squaw Root

Violet

Watercress

URBAN AND
CULTIVATED AREAS

90. ALFALFA
(also Lucerne) *Medicago sativa* (pl. 5f) Family:
Leguminosae

Description: Herbaceous perennial 4-9 dm high with
trifoliate leaves that have tiny teeth around the edges.
Leaflets 1-2.5 cm long and oval in outline. Flowers blue-
violet to purple, 8-12 mm long, resembling a small pea flower.
Fruit is a small coiled pod. Blooms April to October with
fruit following.

Distribution and Habitat: This is the familiar cultivated
hay. It escapes frequently throughout the state along roads,
freeways, and other disturbed areas.

Uses—Past and Present: Alfalfa has been cultivated for
over two thousand years and is probably the first cultivated
forage plant. Native to Southwestern Asia, it was grown by
Persians, Greeks, and Romans. It was introduced early into
China and Europe and reached the United States during the
colonial period. The name Alfalfa comes from an Arabic
word meaning "father of all foods," because that race of
people was convinced of its goodness for themselves as well as
for their horses. The leaf is rich in vitamins A, K, and D,
minerals, and protein. It has been used to encourage the
clotting of blood and in pablum for infants, as well as a
treatment for ulcers. It is much used with Red Clover and
Spearmint or Peppermint as a substitute for pekoe tea and is
an important honey bee plant. The Indians of Utah ground
Alfalfa seeds between stones and cooked the meal as mush or
bread. They boiled young branches for greens.

In recent times, Alfalfa sprouts have become popular as a
salad and sandwich addition. They can be eaten in salads,
steamed, in scrambled eggs, soups, meatloaf, bread, and
stews. Alfalfa tea can be made by drying both leaves and

flowers and steeping a handful of them until the desired strength is reached. They are frequently sprinkled dried and crumbled on cereal. To me, the fresh leaves do not have much to recommend them in flavor, however, their nutritional content makes them a welcome addition to many dishes.

Alfalfa sprouts can be raised at home. Various sprouting jars and plates are used by enthusiasts. The object is to keep the seeds moist but ventilated—not immersed. Probably one of the simplest sprouters is a quart jar with a large mouth, covered with screen or cheesecloth. Alfalfa seeds should be soaked for a few hours in clean water, then placed in the container and set in a dark cupboard. The jar is turned mouth side down at a 45° angle in a bowl to allow ventilation and drainage. Seeds should be rinsed 2 or 3 times a day to prevent souring.

MEAL-IN-AN-OMELETTE

1 Tbsp butter	1/2 tsp salt
or margarine	1/4 tsp pepper
3/4 cup Alfalfa sprouts	1/2 cup grated cheese
2 eggs	
2 Tbsp milk	

Melt butter or margarine in skillet and sauté sprouts until tender. Remove sprouts. Mix remaining ingredients, except cheese, together. Pour this mixture into pan and cook until set, pricking with a fork so that the liquid on top goes to the bottom and gets cooked at the same time. Add sprouts and cheese, fold omelette over, and cook on lowest heat possible, covered, until cheese is melted (about 3 minutes).

ALFALFA TEA

Boil 1 cup of water per serving and pour it over a handful of fresh leaves, or 1 heaping tsp of crushed dried leaves. Cover and let steep until of desired strength. Strain. When mixed half-and-half with mint, this tea is known to be a rich source of vitamins A, D, E, and K, as well as calcium, iron, and manganese.

91. Green AMARANTH

(also Rough Pigweed, Redroot) *Amaranthus retroflexus* (pl. 6a) Family: Amaranthaceae

Description: Annual herb with alternate leaves and small green flowers. Plant is 3-15 dm tall with ovate leaves 3-10 cm long on petioles 1-7 cm long. Flowers occur in dense spikes, 8-20 mm thick, crowded into terminal panicles. Seeds are compressed, rounded, and black. They resemble miniature contact lenses.

Distribution and Habitat: This and numerous other edible species occur in waste places throughout the state.

Uses—Past and Present: In folklore, the flower of one species of amaranth is regarded as the symbol of immortality. The name is taken from a Greek word meaning "incorruptible." The Greeks spread the flowers over graves to demonstrate their belief in the immortality of the soul. The Greek poet Homer mentions that people of Thessaly wore crowns of amaranth at the burial of Achilles.

Indians of Arizona so valued the seeds of amaranth that they cultivated the plant. They parched and ate the seeds whole or ground them into flour. They also used leaves and stems of the young plants as a potherb. Hopi, Pima, Papago, and Havasupai were some of the Yuman tribes using the amaranth. They brought the seed heads home to dry—often under a shelter of willow made just for that purpose. The spikes were beaten with sticks on a hard floor and then winnowed. The Cahuillas used *A. fimbriatus,* known as pigweed, for the same purposes. Indians in the Andes are also reportedly users of the amaranth in significant quantities.

Amaranth greens are quite bland when young and can be prepared alone, as one would spinach, or mixed with stronger flavored greens such as mustard. The best greens are picked before the plant flowers. To gather seeds, pick the green flowers or drying stalks before the seeds shatter out on their own. Lay these on a plastic sheet in the sun or dry them in a slow oven and thresh out by shaking or walking on them. The tiny flower parts may be ground up with the seeds or winnowed out by sifting the seeds through your fingers in a breeze. The seeds are best if roasted (no more than 1/2-in deep in a roasting pan, at 350°, for 1 hour) before grinding. The seeds are relished by Mourning Dove, House Finch, goldfinches, and sparrows.

AMARANTH MUFFINS

1 cup amaranth meal (of seeds or seeds and flower parts)
1/2 cup whole wheat flour
2 tsp baking powder
1/2 tsp salt
3 Tbsp brown sugar or molasses

2 egg yolks
1 1/4 cups milk or
1 1/2 cups buttermilk
1/4 cup melted butter
or margarine
2 egg whites, beaten stiff

Preheat oven to 375° (quick mod.). Mix dry ingredients together and add yolks, milk, and butter or margarine, blending just until moist. Fold in egg whites and bake in greased muffin cups for 20 minutes. Makes about 1 1/2 dozen. This recipe may be thinned with more milk or buttermilk to make pancakes.

SAVORY AMARANTH GREENS

1/2 cup sour cream
1 pkg dry blue cheese
salad dressing mix

1 qt young amaranth
leaves, washed and torn
into bite-sized pieces

Preheat oven to 375° (quick mod.). Combine sour cream and salad dressing mix; fold into amaranth. Pour mixture into a deep casserole dish and bake, uncovered, for 15 minutes or until bubbling hot. Serves 4.

92. Common BARLEY

Hordeum vulgare (fig. 58) Family: Gramineae

Description: Erect annual 6-12 dm tall with leaves 0.5-1.5 cm wide. Often confused with wheat, but barley is distinguished by long awns which are permanently attached to the grain and more rows of grain around the head. Wheat has grains in 2 rows, while barley has 4-6 rows. Both, if found wild, may be treated similarly, however. Grain is produced in summer.

Distribution and Habitat: Found in waste places and fields as an escape from cultivation or planted to retard erosion. When gathering this or other wild grains, care should be taken to assure that the grains are green or golden and not black with fuzzy projections which could be a poisonous fungus known as Ergot.

Uses—Past and Present: Barley was cultivated at the time that writing was invented, and crops of it were known in Egypt as early as 1680 B.C. Egyptians claim that barley was the first of the cereals made use of by man, and they trace its introduction to the goddess Isis. Barley was extensively cultivated in England and appears on the coins of the early Britons. It was frequently used for making beer by many ancient cultures. The seeds of several species of barley were used by Indians in Utah, Nevada, Oregon, and California after its introduction. Aside from making parched-seed flour, some tribes made a coffee substitute from the singed seed coats.

Foragers today consider barley one of the finest grains because of its abundance and size. I have tried several ways to thresh it out and find a modification of the Indian method best. Gather the green or golden grains, dry or roast them, and remove the larger chaff such as stems and leaves. Place about 1/2 cup of grain and attached husks in a wire strainer, sifter, or other porous but noncombustible container, and hold this over an outdoor flame. Shake vigorously to parch and burn off the excess chaff. Some of the grains will pop like popcorn. When all grains are dark or popped, stir and rub off as much charcoal as possible. Transfer these grains to a blender and grind for a coarse meal or grind finely in a flour mill. Using this method, it takes about 30 minutes to prepare enough flour for 1 loaf of bread. If there are coals left from a backyard barbecue, stir them into the grains and the parching may be accomplished without the flame. Every bit of husk does not have to be removed from the grain, for after parching, grinding, and sifting, the husks will not pulverize like the grain and can be sifted out and discarded. This barley

Fig. 58 Common BARLEY
(Hordeum vulgare). 1/2 X.

meal, besides being useful for baked products, may be added to soups or cereals. Some varieties of barley have a loose chaff that falls off easily and can be winnowed out.

Harvested fields of barley are feasts for game birds and songbirds. Ground squirrels also relish the grain.

OLD-FASHIONED BARLEY BREAD

2 cups milk	1 tsp mace
2 cups barley meal or flour	1 pkg active dry yeast
1/2 cup brown sugar	1/2 cup warm water
1 Tbsp salt	5 cups flour
2 Tbsp shortening	(whole wheat or un- bleached white are best)

Scald milk; stir in barley flour, brown sugar, salt, shortening, and mace. Remove from heat and cool to lukewarm. Sprinkle yeast on warm water and stir. Add milk mixture and 2 cups flour to yeast. Beat with electric mixer on medium speed for 2 minutes, scraping bowl occasionally (or beat with a spoon until batter is smooth). Add enough remaining flour, a little at a time, blending first with a spoon and then with the hands, to make a soft dough that leaves the sides of the bowl. Turn onto a floured board and knead until dough is smooth and elastic (8-10 minutes). Place in a lightly greased bowl, turn dough over to grease top, cover with cellophane wrap and then a towel, and let rise in a warm place (such as a gently warmed oven with the door propped open and only the pilot light left on) until doubled (1-1 1/2 hours). Punch down and let rise again until nearly doubled (about 45 minutes). Turn onto floured board and divide in half. Round up to make 2 balls. Cover and let rest 10 minutes. Shape into loaves and place in greased loaf pans. For a pretty look, sprinkle rolled oats on top and press in gently. Let rise until doubled (about 1 1/4 hours). Bake at 375° for 40 minutes. If bread starts to brown too much on top, cover loosely with foil.

BARLEY-RAISIN COOKIES

1/4 cup milk or cream	2 cups barley flour
2 Tbsp oil	1 tsp baking soda

3 Tbsp honey   2 tsp baking powder
1 tsp vanilla   1/2 tsp salt
2 eggs   1/2 cup raisins (soaked
1/2 cup dark brown sugar   in warm water for
                                  a few minutes)

Preheat oven to 375°. Combine liquid ingredients in a bowl. Add sugar. Sift together dry ingredients in another bowl, add raisins, and combine both bowls, just until dry ingredients are moistened. Drop onto greased cookie sheet and bake for 8 minutes. Dates or currants may be substituted for raisins. Makes 2 dozen.

93. Sugar BEET
(also Swiss Chard, Garden Beet) *Beta vulgaris*
(pl. 6b) Family: Chenopodiaceae

Description: Shiny-leaved biennial herb with large root. Leaves are alternate and occur mostly in a basal rosette with long petioles. Upper leaves are reduced and have very short petioles. Stems 3-12 dm tall; leaf blades 1-2 dm long, oval. Flowers are small and greenish in panicled spikes. Base of plant is sometimes red. Blooms July to October or earlier.

Distribution and Habitat: A garden escape that has become established and naturalized throughout much of the state. Especially abundant in low or damp places. Three forms of this plant are cultivated for their roots or leaves. It is related to Spinach.

Uses—Past and Present: Swiss Chard was the Beta of the ancients and Middle Ages. There are many forms of it, and the red, white, and green forms are named from early times; the red by Aristotle, the white and dark green by Theophrastus and Dioscorides. The Sugar Beet was the start of an industry in France in 1811, much promoted by Napoleon I. Today the beets are a major source of sugar in California, and Swiss Chard is found in most supermarkets. The red beet root is also a familiar vegetable. A crude form of molasses may be made by cleaning, crushing, straining, and boiling down the root extract of the Sugar Beet. Roots left to dry will also collect a pocket of molasses in their central cavities. The product most relished by foragers, however, is the leaf.

Similar to Romaine lettuce in texture when young, and tasty even after the plant flowers, the leaves can be gathered most of the year and used fresh or cooked. They maintain their succulence when cooked and thus lend themselves to making "cabbage" rolls. In cooking, Swiss Chard may be substituted into any spinach recipe. In this book, the potherb recipes for Plantain, Amaranth, Cheeseweed, Lamb's Quarters, and Miner's Lettuce may all be interchanged.

WILTED BEET GREEN SALAD

4 slices bacon,
  cooked and crumbled
1 Tbsp bacon drippings
1 hard-cooked egg, diced
1 Tbsp wine vinegar

1 tsp salt
1/4 tsp dry mustard
2 qt beet greens, torn
  into bite-sized pieces

Place all ingredients but greens in a skillet and heat until simmering. Add beet greens, stir briefly, cover, and set off heat for 2 minutes. Stir and serve warm. Serves 4.

BEET MEAL-IN-A-DISH

1 Tbsp oil
1 Tbsp butter
  or margarine
1 lb ground beef
1 small onion, diced
1/2 tsp basil
1/4 tsp marjoram

1/4 tsp oregano
1 tsp salt
1/4 tsp pepper
1 qt chopped
  beet greens
4 eggs, beaten

Heat oil and butter or margarine in a skillet. Add ground beef and cook until browned. Add onion and cook until tender. Stir in seasonings and beet greens and simmer until greens are tender. Add eggs and cook, stirring, until eggs are set. Serves 4-6. For Mock Cabbage Roll, prepare filling without beet greens. Roll 1 Tbsp filling into each slightly boiled or steamed beet leaf. Cover with 2 cups chopped, stewed tomatoes, and bake at 375° for 25 minutes.

The long petioles of the leaves are celerylike and lend themselves very well to celery dishes or fritters like those below.

MOCK ONION RINGS

4 doz beet petioles (6 in or longer)	1/2 tsp salt
1/2 cup plus 1 Tbsp milk	1/2 cup finely chopped onion (dried or fresh)
1 egg	2/3 cup flour
1/2 tsp onion salt	oil for deep frying

Steam or boil beet petioles until just flexible. Drain. Take a flat toothpick and form the petiole into a ring, piercing both ends with the toothpick to hold the ring together at the top. Beat remaining ingredients together. Dip rings into batter and fry in deep fat until golden brown. Serve hot (or make ahead and warm in the oven on a tray with paper towelling for 15 minutes). Makes about 4 dozen. Catsup is a good dip. These are delicious appetizers and may be varied by adding grated cheese or a salad dressing in place of the onion.

94. CHEESEWEED

(also Mallow) *Malva parviflora* (pl. 6c) Family: Malvaceae

Description: Annual, erect, 3-8 dm tall with sparse stellate pubescence. Leaves are roundish and resemble those of a common geranium, 2-8 cm wide on long petioles. Flowers are small, 5-petaled, and resemble a tiny hibiscus. They are white to pink or bluish and occur in axils of leaves. Petals are 4-5 mm long. Fruit is green and disk-shaped, separating upon maturity to numerous 1-seeded segments. Its resemblance to a small, round cheese gives its most common name. Blooms most of the year.

Distribution and Habitat: This and several other species of *Malva* (all edible) may be found in waste places throughout most of the state.

Uses—Past and Present: Mallow was among the plants raised at Delos, Greece, for the temple of Apollo as a symbol of the first nourishment of man. Pliny wrote that a spoonful of mallow would rid one of all diseases. One species of *Malva* is cultivated along the banks of the Nile in Egypt where it is used as a potherb. The young shoots become a salad in France and Italy. Another species is used in China.

The slightly fuzzy forms of Cheeseweed are tasty, but the raw texture is unpleasant to many people. Cooked, however, its flavor shines through, especially if mixed with a sauce that softens the tiny fuzz. The young green (cheese-like) fruits are good raw in salads, minus their enclosing husk. With the husk, they can be cooked into many different dishes much like peas. The ground fruit (and a few leaves for color) can be used for a creamed vegetable soup which resembles pea.

MALVA WITH BACON AND RAISINS

2 qt *Malva* leaves	1/2 cup golden raisins
2 Tbsp bacon fat	3 Tbsp wine vinegar
6 slices bacon, fried crisp	

Boil or steam leaves until just wilted. Brown raisins in bacon fat until plump, adding wine vinegar as they cook. Pour vinegar and raisins over cooked leaves and crumble bacon on top. Serves 4-5.

CHEESEWEED AND BLUE CHEESE

2 qt Cheeseweed leaves, torn into pieces	3/4 pkg dry blue cheese salad dressing mix
1/2 cup sour cream	

Steam or boil leaves until wilted. Blend sour cream and salad dressing mix and stir into leaves. Serve hot. Serves 4.

CREAMED CHEESEWEED SOUP

2 1/2 cups milk	1 small onion, sliced
1 Tbsp flour	2 stalks celery, sliced
2 Tbsp butter or margarine	1 1/2 cups Cheeseweed fruits, husks, and a few leaves
1 tsp salt	1/2 cup chopped raw bacon or smoked ham
dash of pepper	

Put all ingredients except meat in blender and blend until smooth. Add more leaves if not green enough. Pour into

saucepan, add meat, and simmer gently, covered, for 30
minutes. Serves 4.

Other recipes in this book calling for cooked greens may be
used. See Plantain, Amaranth, Lamb's Quarters, Miner's
Lettuce, and Beet.

95. Common CHICKWEED
(also Starwort) *Stellaria media* (fig. 59) Family:
Caryophyllaceae

Description: Low annual with weak threadlike stems 1-4
dm long. A single line of hair runs down one side of the stem
between leaves and changes sides at each pair of leaves.
Leaves are opposite, oval with pointed tips, 1-3 cm long, and
have short petioles or are sessile. Petals of the white flowers
are 4-5 in number and deeply cleft, about 4 mm long. Blooms
February through September.

Distribution and Habitat: A common weed in shady places
throughout most of the state, including fields, gardens, and
lawns.

Uses—Past and Present: Chickweed is widely used raw or
as a potherb in Europe, where it is native. Indians used the
tiny seeds for bread or to thicken soups. It is one of the few
herbs possessing a good copper content. Chickweed was once
sold in the streets to be cooked, used as a poultice for
abscesses and carbuncles, or infused into a tea to comfort
troubled stomachs or to slim down fat people. It can be used
in sandwiches or salads or cooked like asparagus tips. It
shrinks when cooked so a good supply is needed. Its bland
flavor goes well with stronger greens such as Chicory or
mustard. The tender leaves are relished by game birds and
songbirds.

SPICED CHICKWEED

2 lb chickweed	1/2 cup chopped arrow-
1 Tbsp bacon fat	head tubers, cattail
1/2 cup finely chopped	hearts, nut-grass tubers,
onion	or potatoes

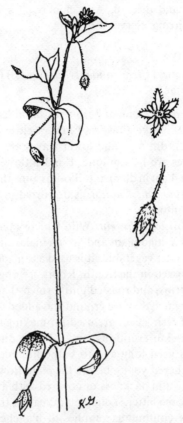

Fig. 59. Common CHICKWEED *(Stellaria media)*. Actual size.

2 Tbsp sugar	4 peppercorns
2 cups water	2 bay leaves
1/2 lb Canadian bacon	1/4 tsp caraway seeds
or ham, diced	salt to taste
5 whole juniper berries	

Clean and drain chickweed. Melt bacon fat in a 3-qt casserole dish over moderate heat. Add onion and cook until tender. Add sugar, water, tubers and meat. Tie seasonings in cheesecloth and drop into casserole. Simmer for 25 minutes or until water is almost evaporated. Remove bag of seasonings and discard. Add chickweed and simmer for 3 minutes, stirring. Serves 4-6.

96. CHICORY
(also Succory) *Cichorium intybus* (fig. 60) Family: Compositae

Description: Perennial herb with basal leaves and bright blue Dandelionlike flowers. Flowers close in midday sun. Plant is 3-10 dm high and lower leaves are lanceolate with teeth. Leaves are 1-2 dm long. Has 1-3 flowers in upper leaf axils, about 4 cm in diameter. Blooms June through October.

Distribution and Habitat: Naturalized in waste places in much of California.

Uses—Past and Present: Wild Chicory has been used for centuries as a salad plant and grown in darkness to provide a highly esteemed vegetable. It is used as an additive to coffee, especially prized in the South. Roots are dug up, sliced into long, thin strips, and roasted until brown. This takes about 1 hour, after which they are ground and added to or substituted for coffee. Fresh leaves are used for making teas and lotions for bathing skin eruptions, and a liquid distilled from the flowers was used long ago to bathe inflamed eyes. Leaves must be gathered young before the plant flowers for the best eating. Grown in darkness or covered with a can, the leaves will not become bitter. Young plants can be transplanted to a cellar where continuous supplies of blanched, tasty greens can be obtained. Like its cousin, the Dandelion, it has a crown just on top of the root (where new leaves emerge) which is also used in salads. Some nurseries carry seed of an improved form of this species.

97. DANDELION
Taraxacum officinale (fig. 61) Family: Compositae

Description: Perennial with leaves in a basal rosette. Leaves toothed, flowers yellow, fruit bearing numerous pale capillary plumes in a globular shape. The familiar lawn weed. Blooms most of the year.

Fig. 60. CHICORY *(Cichorium intybus).* 1/2 X.

Distribution and Habitat: Dandelions are found in most localities where it is damp throughout the state.

Uses—Past and Present: The common Dandelion was eaten and used in American Indian medicine soon after its introduction into the northeastern United States. A tea of the roots was used for heartburn by the Pillager Ojibwas, while the Mohegans and other tribes drank it for its tonic properties. The Iroquois preferred the boiled leaves with fatty meats, as did the Papago and Cahuilla tribes of Arizona and California, respectively. Among the Tewas, the leaves were used to treat a fracture, ground and mixed to a paste and spread on the injured part. A tonic for heart trouble was made from the blossoms by some tribes. Deer, moose, elk, bears, rodents, and many birds, including grouse and pheasant, consume the plants. The dried root was listed in the U.S. pharmacopoeia from 1831 to 1926.

Fig. 61. DANDELION *(Taraxacum officinale)*. 3/4 X.

The Dandelion has saved many people from starvation and is relished by many others. It contains healthy amounts of vitamins A, B, and C and calcium, sodium, and potassium. Generally, the products are best before the plant flowers, at which time they become too bitter. Young, peeled roots, greens, crowns (root-leaf junction), and flowers are used in various ways. The best greens are gathered not in closely mowed lawns, but in flower beds or other areas where they have been allowed to grow. They are used raw in salads or cooked. A popular French salad is made by mixing the leaves with a light dressing of olive oil and a squeeze of lemon. This is garnished with finely chopped chives, parsley, garlic, or borage. Another variation adds bacon or pork. Leaves are steamed or boiled until tender and merely served with butter or spices. Before the flower emerges, the crown of the plant is white and tender and makes a fine vegetable raw or cooked. The developing buds resemble artichoke in flavor. The fresh roots are sometimes eaten as a vegetable; peeled, sliced, and boiled in 2 waters with a pinch of soda added, they are served with salt and butter.

To make a coffee substitute of the roots, clean, peel, and dry them and roast until brown throughout. Then grind and use as for coffee or tea.

DANDELION WINE

1 gal water	10 whole cloves
3 qt Dandelion flowers	3 lb sugar
2 lemons	1 pkg active dry yeast
2 oranges	1 tsp yeast nutrient (if available)

For best results, all utensils must be clean and sterilized in boiling water. Use only flowers, not stems or leaves. Discard as much of the green parts as possible. Cover flowers with 3 qt boiling water, cover, and let steep for 3 days. Add rind from lemons and oranges (no white pith) and cloves. Boil for 30 minutes, covered. Strain the liquid into a crock or glass container and add the juice from the lemons and oranges. Dissolve as much of the sugar as possible in 1 qt water by boiling and stirring. Add half of this sugared water to the crock. Cover and store remaining sugared water at room temperature. When cool, add yeast and yeast nutrient, cover and ferment for 3 days in a warm place. Stir in 1/2 remaining sugared water and let ferment for 1 more day. Decant into bottles and fill to within 2 in of top of the neck of the bottle, using remaining sugared water or plain water. Place an air lock or a cotton plug covered with a piece of loosely tied plastic over the neck of the bottle. The object is to let gases through but not let flies touch the neck of the bottle, for they carry vinegar-causing bacteria on their legs. Let this ferment in the dark for 3 more weeks until all fermentation has ceased and cap. Age 6 months before drinking.

DANDELION GREEN SALAD

1/2 gal Dandelion greens (picked before flower has arisen)	2 strips bacon, fried crisp
Dandelion flower buds, unopened (if available)	3 Tbsp vinegar-and-oil dressing

Wash greens. Crumble bacon over greens and buds and add dressing, tossing lightly. Serves 4-5. Other greens may be added for variation.

A cream of Dandelion soup may be made by following the recipe for Creamed Cheeseweed Soup.

98. Red-stem FILAREE
(also Storksbill, Clocks, Pin Grass) *Erodium cicutarium* (fig. 62) Family: Geraniaceae

Description: Small annual herb with leaves that form a rosette at the base. Leaves are lacy in outline due to many dissections of the blade. Stems are slender and weak, 1-5 dm long with fine hairs. Flowers are rose-purple with 5 petals, 5-7 mm long, each with 2 spots at the base. There are 5 stamens and fruit consists of a needlelike structure at the base of which are 5 seeds. When mature, the fruit can be agitated between the fingers and it will release the seeds, each with an attached needle. This structure begins at once to corkscrew around, as it would to drive the seed into the ground—thus the common name Clocks.

Distribution and Habitat: This and other species of *Erodium* are found throughout the state in deserts as well as wet and disturbed areas below 6000 ft (1800 m). Other species are edible, but this has the best flavor in my opinion.

Uses—Past and Present: Shortly after its introduction, the American Indians quickly learned to use this plant for food. Blackfeet, Shoshone, and Cahuilla Indians gathered the young plant and cooked it or ate it raw. Though the lacy leaves are of a somewhat unfamiliar texture, the cooked greens are similar to spinach and may be used in recipes for the leaves of beet, plantain, Sow Thistle, and amaranth. Quail, finches, ground squirrels, and kangaroo rats are among the animals that eat the plant.

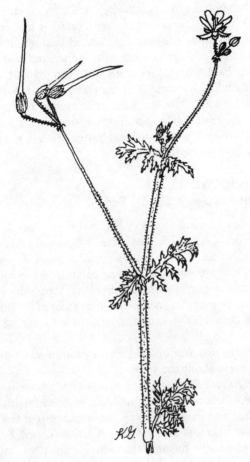

Fig. 62. Red-stem FILAREE *(Erodium cicutarium).* Actual size.

FILAREE FRITTATA

6 Tbsp butter
  or margarine
1/2 cup chopped onion
1/2 cup chopped celery
1 clove garlic, minced
1/4 tsp basil leaves
1/8 tsp oregano

1 qt filaree leaves,
  packed
6 eggs
3 Tbsp grated
  Parmesan cheese
1/4 tsp salt
1/8 tsp pepper
3 Tbsp cold water

Melt 3 Tbsp butter or margarine in a skillet and sauté onion, celery, garlic, basil, and oregano until celery is tender. Add filaree and continue cooking until heated through. Remove from heat. Combine eggs, cheese, salt, pepper, and cold water in a bowl. Beat with a fork until well blended and stir in vegetables and seasonings. Melt remaining 3 Tbsp butter or margarine in skillet over low heat, pour egg mixture in, and cook slowly, lifting sides and bottom until it sets. When bottom is browned, place under preheated broiler until top is browned. Fold omelette style. Serves 4.

99. HOREHOUND
Marrubium vulgare (fig. 63) Family: Labiatae

Description: Perennial, wooly herb with bitter sap. Leaves wrinkled, toothed, and petioled. Plant 2-6 dm tall, leaves 1.5-4 cm long, white, wooly. Leaves occur in pairs on square stems. Flowers are small, white, and occur in dense whorls. Corolla has 2 major lips, the upper with 2 lobes, the lower with 3. Four stamens are included in the throat of the flower. Fruit is a cluster of 4 small nutlets. Blooms spring and summer.

Distribution and Habitat: Common weed in waste places and fields throughout the state.

Uses—Past and Present: Since ancient times, both abroad and in North America, Horehound has been used to make a candy prized for its soothing (if somewhat bitter tasting) effect on sore throats and coughs. A tea was used by the Cahuilla Indians for flushing the kidneys. The plant can be boiled or dried without losing its flavor, which is unusual for a member of the Mint family. One cup of fresh leaves or 1/4 cup dried leaves boiled in 2 cups of water for 10 minutes will make a strong concentrate. This concentrate can then be diluted with 2 parts water to 1 part concentrate for a tea. One part concentrate may also be added to 2 parts sugar or honey and a pinch of cream of tartar, brought to hard crack (290°), and poured into a buttered plate for the old-fashioned cough drop candy. A bit of lemon added at the last minute improves the flavor. A cough syrup can be made of 1 part concentrate and 2 parts honey.

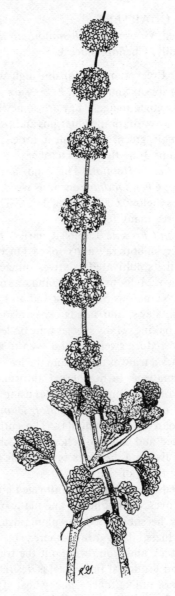

Fig. 63. HOREHOUND *(Marrubium vulgare)*. 1/2 X.

100. LAMB'S QUARTERS
(also Pigweed, Goosefoot) *Chenopodium album*
(pl. 6d) Family: Chenopodiaceae

Description: Erect annual 2-20 dm high with a whitish meal on young shoots and leaves. Leaves are alternate, pale green, and red-veined in age. They are somewhat triangular in outline with wavy or toothed margins and 1-5 cm in length. Flowers are small, green, and occur in little spikes in upper leaf axils. Blooms June through October.

Distribution and Habitat: This and numerous other species of edible *Chenopodium* occur as weeds in waste and disturbed areas below 6000 ft (1800 m). They are quite variable and have many forms.

Uses—Past and Present: Lamb's Quarters has long been used as a wild green both raw and cooked. Once very popular in Europe until its cultivated relative, spinach, came into being. The town of Melbourne in England was named after an old goosefoot. Numerous Southwest Indian tribes used the leaves in soups, stews, and salads. They also gathered the black seeds for baking. Many species were boiled and applied as a poultice to reduce swellings and even to soothe aching teeth. Sometimes a tea was made from the leaves for use as a wash for rheumatism. In northern California, the Pomo, Yuki, and Miwok were among the tribes using the leaves for food. The root of one species. *C. californicum,* was ground and used as soap by the Cahuillas. The entire plant was sometimes boiled and eaten to relieve stomachache. Napoleon reportedly lived on the black bread made from seeds of pigweed during times of scarcity.

Today the seeds are devoured by informed humans as well as Mourning Doves, quail, finches, and many other animals. The seeds may be substituted for plantain and amaranth in baked products. They are gathered by rubbing the flowers into a sack and winnowing out the trash. They are then roasted and ground if finer meal is desired. The flavor they impart is not unlike that of buckwheat. The leaves are much sought today by foragers because of the pleasant taste and high vitamin and mineral content.

GOOSEFOOT GREENS WITH YOGHURT DRESSING

1 cup unflavored yoghurt
2 Tbsp olive oil
2 Tbsp lemon juice
1/2 tsp salt
1/2 tsp sugar

1/4 tsp pepper
1 small clove garlic,
  crushed
1 Tbsp finely chopped
  parsley or mint
Goosefoot greens

Blend together all ingredients except Goosefoot greens. Boil or steam greens until tender (or use fresh). Add about 1 Tbsp dressing to each serving of greens. Makes 1 1/4 cups (about 8 servings).

NAVAJO GRIDDLE CAKES

1 cup whole wheat flour
1 cup roasted and ground
  Lamb's Quarter seeds
1/2 tsp salt
1 tsp baking powder

2 Tbsp sugar or honey
1 egg
2 cups plus 2 Tbsp milk
3 Tbsp melted butter
  or bacon fat

Combine dry ingredients. Beat in liquid ingredients until smooth. Bake on an ungreased hot griddle as you would pancakes, turning when bubbles appear. Serve with butter and hot maple syrup or Manzanita Jelly.

101. Black MUSTARD
Brassica nigra (pl. 6e) Family: Cruciferae

Description: Erect annual, branched above, 0.5-2.5 m high. Basal leaves deeply cut, those of the stem toothed or smooth. Lower leaves 1-2 dm long with large terminal lobes and a few small lateral ones. Flowers showy, yellow, and in elongated racemes. There are 4 petals, 7-8 mm long, and 6 stamens, 2 shorter than the other 4. Fruit is an elongated, slender pod 1-2 cm long in maturity, with a beak at the tip which is empty of seeds 1-3 mm long. Flowers April through July.

Distribution and Habitat: Common on dry hillsides and waste places throughout the state. Many other edible species of this genus exist, this being the most widespread and flavorful. They are so widespread partly because the forest service planted them on burns to retard erosion, and partly because the mission fathers used to spread the seeds to mark the trail from one mission to another along the coast.

Uses—Past and Present: This is the mustard of the ancients and is cultivated in Europe and England. Although the young plants were originally used as a potherb, today they are grown mainly for their seed, from which the commercial mustard is derived. Many cultivated crops are in the Mustard Family, including cabbage, cauliflower, radish, broccoli, and brussel sprouts. There are numerous species of Brassica that have gone wild, and all furnish good food. The best leaves to use are the lower ones. Though sometimes fuzzy, they mix well with other greens or in soup. Because of the strong flavor, they should be cooked for about 20 minutes in water and served with butter or a seasoned, mild dressing. The greens become quite strong when warm weather sets in, so they are best gathered young. They are high in vitamins, especially A, B, and C.

The unopened flower buds can be eaten like broccoli—boiled in salted water for just a few minutes. The yellow flowers give rise to pods that contain several seeds. Collecting these pods and allowing them to dry, one can obtain the seeds. To make prepared mustard from the seeds, grind them in a food chopper or mill and roast in an oven until browned. Mix this flour half-and-half with commercial powdered mustard, moistening with a mixture of half vinegar and half water. Mustard plasters can also be made by using the ground seeds half-and-half with flour and water.

Doves, pheasants, finches, larks, and nuthatches are among the birds that search out the seeds. Ground squirrels and deer eat the plants.

DELUXE CREAM OF MUSTARD SOUP

3 cups mustard greens, washed and torn into small pieces
6 cups milk

1 cup mashed potato flakes or mashed potatoes
5 Tbsp butter or margarine
2 Tbsp flour

1 Tbsp salt
1 cup sour cream

2 cups grated Monterey
jack cheese

In a blender, purée greens in a little of the milk. Pour into a saucepan and add rest of milk, potatoes, butter or margarine, flour, salt, and cheese. Simmer gently until cheese is melted (about 10 minutes.) Remove from heat, stir in sour cream, and reheat gently until smooth. Freeze and reheat gently, if desired. Serves 8.

CALIFORNIA MUSTARD DIP

1 cup sour cream
1/2 tsp salt

2 Tbsp raw mustard leaves,
finely chopped and rinsed

Mix all ingredients together and refrigerate for several hours before serving. A convenient way to chop the leaves is to place them and a little water in a blender, purée for a few seconds, and strain through a tea strainer. Measure after straining. Serve with chips or crackers.

MUSTARD AND COLLARD GREENS

2 smoked ham hocks
    or 1 cup chopped ham
    or bacon
1 gal mustard greens

3 qt collard greens
2 tsp sugar
1 Tbsp salt
1/4 tsp pepper

Place meat in large saucepan, cover with water, and simmer for 2 hours. Tear greens into large pieces without stems, rinse several times in warm salted water, and drain well. Stir sugar, salt, and pepper into water with meat. Add enough greens to fill pot, cover, and when greens shrink, add more, stirring occasionally until all are tender (may take more than 1 hour). Serves 8.

DANISH MUSTARD

2/3 cup mustard powder
1/2 cup dark brown sugar
pinch of salt
1/4 cup cider vinegar

1/4 cup oil
1 tsp Worcestershire sauce
1 tsp lemon juice

Beat together all ingredients and place in a tightly capped jar in the refrigerator for a few days before using. Makes 3/4 cup of a sweet, hot mustard.

102. Wild OAT
Avena fatua (fig. 64) Family: Gramineae

Description: Low to tall grass, 1-2 m high. Flowering structure a loose, open panicle with horizontal branchlets. Greenish flowers have a long, stiff awn which is bent and twisted below, 3-4 cm long. Blooms April through June.

Distribution and Habitat: This and two other species of Oat are found as weeds in waste fields and disturbed areas throughout the state. Care should be taken to gather only green or golden grains, not any which have been blackened by a fuzzy mold, which may be the poisonous Ergot.

Uses—Past and Present: The Wild Oat possibly introduced by the Spaniards, is now spread over the entire country. It was gathered by numerous California Indian tribes, including the Pomo, Cahuilla, and Miwok. They crushed the seeds lightly to loosen the chaff, winnowed out the trash, and ground the parched seeds into meal. Waterbirds and songbirds frequently feed on the grains. Oats should be prepared in the same way as barley, described elsewhere in this book, and may be used in the same recipes.

OAT BREAD

1 cup quick-cooking rolled oats
1 cup nonfat dry milk
2 1/2 cups boiling water
2 pkg active dry yeast
1/2 cup warm water
3 Tbsp oil

1/2 cup molasses, brown sugar, or honey
1 Tbsp plus 1 tsp salt
1 1/4 cups Wild Oat flour
2 1/2 cups whole wheat flour
3 1/2 cups unbleached white flour

In a large bowl, mix rolled oats, dry milk, and boiling water. Cool to lukewarm. Soften yeast in 1/2 cup warm water for 5 minutes. Stir into oat mixture with oil, molasses, and salt.

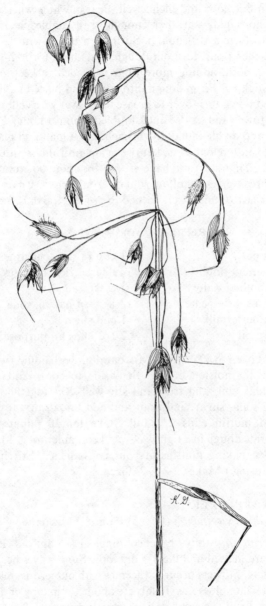

Fig. 64. Wild OAT *(Avena fatua).* 1/2 X.

Beat in oat flour and blend well. Beat in whole wheat flour until thoroughly wet, then enough of the unbleached white flour to form a stiff dough. Sprinkle a board with 1/2 cup unbleached flour, turn out dough, and knead for 15 minutes until smooth, adding more flour if needed to keep dough from sticking. Form dough into a ball and place in a lightly greased bowl, turn over to grease top, cover with cellophane and a towel, and let rise until doubled (about 1 hour). Punch down and divide in half, form each into a loaf, and place in greased 5x9-in loaf pans. Let rise again until almost doubled, (about 30 minutes) and bake at 350° for about 50 minutes or until bread gives a hollow sound when tapped. Cover with foil for last 20 minutes if needed to prevent too dark a crust.

REFRIGERATOR MUFFINS

3 cups oat (or barley) flour	1 cup raisins, currants, elderberries, or dates
1 cup boiling water	1/2 tsp salt
2 eggs, slightly beaten	2 1/2 tsp baking soda
2 cups buttermilk	1 cup sugar
1/2 cup oil	2 1/2 cups all purpose flour

Preheat oven to 425° (hot). Mix oat flour with boiling water, stirring to moisten evenly. Set aside to cool. Add eggs, buttermilk, oil, and fruit and stir well. Sift together salt, baking soda, sugar, and flour and add to oat mixture. Fill greased muffin cups 2/3 full. Bake for 20 minutes (or refrigerate dough in a tightly covered container for as long as 2 weeks, baking muffins at your convenience. Stir batter before using.) Makes 2-2 1/2 dozen.

103. Wild PEPPERGRASS

Lepidium virginicum (fig. 65) Family: Cruciferae

Description: Many-branched annual 1.5-6 dm tall. Basal leaves are incised and 0.5-1.5 dm long. Stem leaves become lanceolate. Flowers occur in racemes with numerous small 4-petaled white blossoms. Petals are about 1 mm long and 4 in number. Fruit is a small pod, nearly heart-shaped, 2.5-4 mm

Fig. 65. Wild PEPPERGRASS *(Lepidium virginicum).* 1/2 X.

long. The most common variety of this variable plant has small amounts of hair on stems. Blooms in spring and summer.

Distribution and Habitat: Widespread in waste places, along roadsides below 7000 ft (2100 m). Numerous other species and forms of peppergrass occur in other habitats. *Descurainea pinnata,* Tansy Mustard, was also used in this manner and bears the same common name. It too is a member of the Mustard Family but occurs primarily in drier areas such as deserts.

Uses—Past and Present: The small seeds were gathered by

many Indian tribes and parched by stirring in a basket with live coals. They were then ground to a flour and made into mush with other seeds. Mexicans use the seeds of a related plant as poultices for wounds. The Cahuillas used a desert species, *L. nitidum,* for a hair treatment. The leaves were boiled and allowed to steep until a brown-colored water was obtained. This was then used to wash the hair and was said to keep the scalp clean and prevent baldness. *L. flavum,* another desert species, has a flavor similar to Watercress. The leaves of these and other species were also eaten roasted or as a potherb. It sometimes takes two or three changes of water to lessen the strong taste. For recipes, see Black Mustard.

104. Common PLANTAIN
(also Goosetongue) *Plantago major* (pl. 6f)
Family: Plantaginaceae

Description: Perennial, stemless herb with thick, broadly elliptic leaves. The leaf is 0.5-1.5 dm long with several conspicuous nerves converging at the base and apex. Petioles are winged. Flowering spikes are 0.5-4 dm high, erect, with small 4-petaled blossoms that are followed by numerous small brown capsules. Blooms April through September.

Distribution and Habitat: Introduced weed of damp places and lawns throughout the state. Numerous other forms and species occur, but the ribbing in the leaves and characteristic fruiting spikes make them easy to recognize. A narrow-leaved variety, *P. lanceolata,* is more tender.

Uses—Past and Present: The history of the seeds as food can be traced back to the Egyptians, for seeds have been found in the tombs of pharoahs, and the leaves were used as a potherb in China. The American Indians called it White Man's Foot in reference to its growing wherever white men set foot. The Shoshone Indians heated the leaves and used them for a wound dressing, while the settlers used them as an antidote for bites of venomous reptiles and insects. The seeds were used as a worm remedy.

Young leaves are today used as a potherb or in salads. They become tough as they mature, but may still be cooked if chopped. A handful of leaves can be steeped for a half hour to make a cup of tea. Seeds are often sought for caged-bird food, parched or ground into meal, or roasted. The whole seeds have a laxative effect if eaten raw. Try young plantain leaves in salads dressed with vinegar and oil or other tart dressing (see recipe for Beet), or cook and make into a soup (see recipes for Cheeseweed and Black Mustard).

STUFFED PLANTAIN

1 lb plantain leaves
  (1 qt, tightly packed)

Pull strings off main ribs first. Then parboil plantain in salted water until tender.

Sauce:

1 Tbsp red wine vinegar	1/2 cup water
1/2 cup red wine	1/2 cup catsup
1 Tbsp sugar	

Filling:

1 lb sausage	2 cups wild rice with
1/2 onion, finely chopped	mushrooms (prepared
1 cup shredded	according to package)
cheddar cheese	1 egg, slightly beaten

Preheat oven to 400° (hot). Brown sausage in pan, drain, and add onion. Cook until onion is tender. Add remaining ingredients and simmer for 5 minutes. Place about 1/4 cup filling in each plantain leaf, roll leaf around it, and place seam side down in baking dish. Mix all sauce ingredients together and pour over stuffed plantain. Cover and bake 20 minutes. With narrow-leaved varieties of plantain, this may all be layered and baked as a casserole. Serves 6.

CREAMED PLANTAIN SOUP

1 1/2 lb plantain leaves	3 Tbsp butter
3 cups water	or margarine
salt to taste	1/4 cup flour

6 cups chicken broth   pepper to taste
1/2 cup heavy cream   3 Tbsp fresh lime
4 egg yolks   or lemon juice
1/8 tsp nutmeg

Remove toughest ribs from plantain leaves, boil in salted water until tender, and drain. Chop fine. There should be 1/2 to 1 cup. Melt butter or margarine and add flour, stirring until blended. Add broth, stirring rapidly, and simmer 30 minutes. Blend the cream with egg yolks. Remove soup from heat and stir in egg yolks and cream. Add plantain and reheat to a gentle simmer, but do not boil. Add remaining ingredients and serve hot or cold. Serves 6-8.

105. PURSLANE
(also Verdolaga, Pusley) *Portulaca oleracea*
(pl. 7a) Family: Portulaceae

Description: Fleshy herb with pinkish stems 1-2 dm long, spreading out from the center along the ground. Leaves are alternate or partly opposite, paddle-shaped, 0.5-2.5 cm long. Flowers are small, yellow, 5-petaled, and occur in angles of stems. There are only 2 sepals. Black seeds occur in a small capsule. Flowers are produced mostly in summer.

Distribution and Habitat: Introduced weed of cultivated fields and lawns throughout the state. It is said that between Canada and Mexico there is probably not one town where it cannot be found.

Uses—Past and Present: Purslane is thought to be native to India, where it was used two thousand years ago. Pliny the Elder wrote 1900 years ago that it was a good medicine to take for fevers. It was once valued for its power to remove magic spells that were cast upon a person or his cattle. The plant was scattered around the bed as protection from evil spirits and was also considered to be a sure cure for the striking of lightning or gunpowder. Medicinally, it was used for many ailments, principally fevers, inflammations, coughs, insomnia, and eye problems.

Purslane is found today in Mexican markets, where it is

known as Verdolaga. Its succulence makes it a welcome addition to gumbos, soups, casseroles, and salads. Stems are often made into pickles. I pinch just the tender tips of the stems and leave the plant to continue producing them for later use. Seeds are gathered by placing plants on plastic sheeting and thrashing them out of their capsules. They are then ground up and make a good buckwheat flour substitute. Birds and rodents also relish these tiny black seeds, while fur and game animals browse the leaves and stems.

PICKLED PURSLANE STEMS

1 qt. Purslane stems
  (or leaves)
1 cup dark vinegar

1 cup sugar
1 Tbsp pickling spices
  tied in cheesecloth

Wash stems and cut into 2-in pieces. Put in cold water with 1/4 cup salt to each qt water and let stand 4-5 hours. Bring vinegar, sugar, and spices to a boil. Drop in Purslane and bring to boil again. Simmer for 10 minutes. Pack stems into sterilized jars and pour over enough syrup to cover. Seal. Let stand 2 weeks. Makes 1 quart. A faster way to pickle if you want to use them right away is to just add a pinch of pickling spices to wine vinegar, place raw stems into this, and marinate for several days. Use chilled.

VERDOLAGAS CON QUESO

1 qt Purslane stems
  and leaves
1 medium onion, chopped
1 Tbsp bacon fat

3 slices bacon, fried
  and chopped
1/4 lb Monterey jack
  cheese, shredded
salt and pepper to taste

Break off heaviest stems of Purslane and discard or pickle. Rinse remainder and cook in boiling salted water for about 5 minutes or until just tender. Drain off water and chop Purslane. Cook onion in bacon fat until tender and add to Purslane in a saucepan. Add chopped bacon, cheese and salt and pepper and heat until cheese melts. Serves 4-6.

PURSLANE AND SHRIMP VINAIGRETTE SALAD

1 qt Purslane	1/8 tsp pepper
leaves and tips	1/4 cup white wine vinegar
1 hard-cooked egg	1/2 cup olive oil
3 green onions, minced	1/4 lb tiny cooked shrimp,
1 Tbsp minced parsley	shelled
1/4 tsp salt	

Rinse and chill Purslane. Mash egg and combine with green onion, (include tops). Combine remaining ingredients and toss with Purslane and egg and onion mixture. Serves 6.

106. Wild RADISH

Raphanus sativus (pl. 7b) Family: Cruciferae

Description: Freely branched herb 3-12 dm tall with lyre-shaped leaves at the base, 1-2 dm long with pinnate divisions and large, rounded terminal lobes. Flowers have 4 petals, white with rose or purple veins, sometimes yellowish. Petals 1.5-2 cm long. There are 6 stamens. Fruit is a rounded podlike structure 2-3 cm long. Blooms February through July.

Distribution and Habitat: Common weed in waste places and fields throughout the state. The garden radish is a cultivated form of this same species.

Uses—Past and Present: The radish originated in China, where it has many forms. It was also cultivated extensively in Egypt during the time of the pharoahs. It grows freely around the Mediterranean region, especially Greece. Ancient Greeks, in giving offerings to Apollo, presented turnips in lead containers, beets in silver, but radishes in beaten gold. In some countries the leaves and pods are favored foods, rather than the root. The use of radish as a salad plant are too well known to need discussion, however, this wild form often bears a root that is too woody to use in the traditional ways. The leaves may be made into soup or dip in the same way Black Mustard is used, and the flowers may be tossed in salads or eaten alone as a snack. The fruiting pods, resembling a small green bean, make tasty salad additions if they are gathered before the seeds harden and the pods dry out. The taste is almost identical to the garden radish.

107. Australian SALTBUSH
Atriplex semibaccata (pl. 7c) Family: Chenopodiaceae

Description: A low, prostrate perennial with many branches. Many leaves are oblong, 1-3 cm long. Male flowers are in small terminal bunches, while the female are in axils of leaves. Fruit is a fleshy red berry 3.5-5 mm long, rhombic in shape. There is a dusting of minute scales on surface of leaves.

Distribution and Habitat: A common bush of saline and waste areas found throughout the state from San Luis Obispo Co. to Baja California. Numerous other members of the genus are also useful. See, for instance, Desert Holly and Saltbush.

Uses—Past and Present: The red berries of this species of Saltbush make nice snacks or trim for salads. They resemble a tomato in taste. Indians used the seeds of many native species of saltbush for flour. The leaves and young shoots of the native species are often used as potherbs and are quite agreeable with meat. The Zuni Indians ground roots and blossoms of *A. canescens* and moistened them with saliva for use on ant bites. *A. tularensis* and *A. vallicola* of northern California are on the rare and endangered plant list for the state.

108. SHAGGY MANE
(also Inky Cap) *Coprinus comatus* (fig. 66) Class: Basidiomycetes

Description: Among the easiest of all the gilled fungi to recognize, the Shaggy Mane has a dark spore deposit when mature. Gills dissolve or become inky as cap matures. Cap is 4-10 cm high and 3-4 cm thick, ellipsoid to cylindric when young, gradually expanding in age. Surface is scaly with tips of scales yellowish-brown and the remainder whitish. Where the cap separates from the stipe, a loose ephemeral ring

Fig. 66. SHAGGY MANE *(Coprinus comatus).* 1/2 X.

(annulus) is left. The stipe is hollow with central strand of white fibers. As with many wild plants, mild poisoning has been reported, probably due to individual allergies. This is considered by many to be one of the most delectable of all mushrooms. Occurs throughout winter.

Distribution and Habitat: Shaggy Mane is found in clusters throughout urban areas of the state, particularly along roadsides and in lawns and gardens. It has even been reported pushing up through the paving of tennis courts. It is found growing in the hardest of soils.

Uses—Past and Present: Mushrooms were not heavily used by the Indians of America in general, although a few were prized by the Miwoks and Cahuillas. Among the Tewas it was believed that a stick must be laid across the top of the kettle containing cooked mushrooms or the person eating them would be afflicted with a poor memory.

Unless one gets young specimens, the lower edges of the gills of Shaggy Mane will have begun to turn dark, but this can be cut off and discarded. They can be preserved by slicing and drying or sautéed and frozen. They should be cooked soon after picking, since the process of autodigestion which turns the cap from an edible white to an inedible black continues after picking—often even in the refrigerator. As in all the fungi, care must be taken to insure that correct identification has been made before use. Other recipes will be found under Morel and Puffball.

MUSHROOMED CHICKEN IN WINE SAUCE

4 chicken breasts
1 clove garlic, crushed
2 cups sliced fresh mush-
  rooms (or dried ones
  soaked for a few minutes)
1/2 tsp seasoned salt
1/8 tsp pepper

1/2 medium onion, sliced
3 Tbsp butter
1 cup chicken broth
  or soup
1/2 cup white wine
1 tsp tarragon
1/2 cup ham, chopped

Rub chicken with garlic and broil or roast until tender. Combine remaining ingredients in a saucepan and simmer 4 minutes until mushrooms and onions are tender. Add chicken and simmer, covered, for 5 minutes. Serve with rice. Serves 4 generously. The sauce may be thickened by adding a little cornstarch. The alcohol in the wine evaporates, leaving only the delicious flavor.

109. SHEEP SORREL
(also Sour Dock) *Rumex acetosella* (fig. 67)
Family: Polygonaceae

Description: Perennial herb with tufted stems 1-4 dm tall. Leaves lanceolate, 2-6 cm long, the lower ones with an arrowhead shape and the petioles longer than the blades. Flowers are in nodding panicles, yellowish but reddish in age. Fruit a small triangled reddish nutlet. Flowers are produced March through August.

Distribution and Habitat: This and another species, *R. angiocarpus,* are both known as Sheep Sorrel. They are common weeds in damp places throughout the state. Numerous other edible species are called dock.

Uses—Past and Present: These Sour Docks are highly esteemed by many peoples around the world as an additive to

Fig. 67. SHEEP SORREL
(Rumex acetosella). 3/4 X.

salads, soups, and bread. They are related to rhubarb and the stems of some species are used in the same way (see Canaigre). They are highly prized in French gardens and in India are used for omelettes. The Miwoks of California pulverized the leaves of this species, moistened them with water, and ate them with salt. The seeds are a common item in the diet of ground-feeding birds, while Mule Deer eat the leaves.

A lemonadelike drink can be made by simmering the leaves in water for 20 minutes. Cottage cheese with chopped Sheep Sorrel is an excellent salad. The leaves are also tasty additions to coleslaw, sandwiches, vegetarian drinks, and fish dishes. If cooked, they resemble spinach with lemon. To reduce the sour taste, they may be cooked in 2 changes of water. The leaves of many species of dock are useful as potherbs.

SORREL SALAD

6 slices bacon
3 Tbsp salad oil
2 Tbsp sugar
2 Tbsp catsup
2 Tbsp red wine
  vinegar
2 Tbsp minced green
  onion

2 Tbsp minced parsley
  flakes (optional)
1/2 tsp Worcestershire
  sauce
dash of garlic salt
2 qt sorrel leaves, torn
  in bite-sized pieces
salt and pepper to taste
2 hard-cooked eggs, sliced

Fry the bacon until crisp, drain, crumble, and set aside. Combine the oil, sugar, catsup, vinegar, onion, parsley, Worcestershire, and garlic salt. Stir until well blended. Pour over sorrel, add bacon, and toss together. Season to taste with salt and pepper and garnish with egg slices. Makes 6-8 servings. Dressing may be made ahead if covered and refrigerated.

SHEEP SORREL PIE

1 qt Sheep Sorrel leaves,
  shredded

2 cups water
6 egg yolks

3/4 cup sugar   1 Tbsp powdered sugar
6 egg whites   1 9-in unbaked pie shell

In saucepan, bring water to boil and add sorrel leaves. Cook for 3 minutes, remove from heat, and steep 1/2 hour. Beat egg yolks until creamy and add 1/2 cup sugar, beating until thick. Stir in 1/2 cup strained juice of Sheep Sorrel. Cook over low heat for 5 minutes or until thickened. Do not allow to boil. Remove from heat and cool. Beat egg whites until frothy, slowly adding 1/4 cup sugar until stiff peaks are formed. Mix about 1/3 of this into egg yolk mixture, then gently fold in the rest. Set aside to cool to room temperature, then chill in refrigerator. Pour into pie shell and bake at 400° until firm and brown (about 30 minutes). Dust with powdered sugar before serving cold.

110. SHEPHERD'S PURSE
(also Mother's Heart) *Capsella bursa-pastoris*
(fig. 68) Family: Cruciferae

Description: Erect annual with forked, short hairs. Plant is 2-5 dm tall and usually branched at base. Leaves form a rosette and are 3-8 cm long at base of plant. They are incised rather like Dandelion. Leaves along the stem are lanceolate with lobes at the base and no petioles. Flowers are small, white, with 4 petals, 2 mm long, and 6 stamens. Fruit is a heart-shaped little capsule with numerous tiny seeds. Blooms most of the year.

Distribution and Habitat: A common weed below 7000 ft (2100 m) throughout the state.

Uses—Past and Present: Shepherd's Purse was used in China as a potherb and reportedly cultivated for this same purpose at one time in Philadelphia. The Cahuillas gathered the leaves from January to June for greens and the seeds during the summer. It has also been reported that they used the plant to make a tea for the treatment of dysentery. The leaves should be gathered before the plant flowers and are even better when blanched by turning a pot over the plant or growing them in a basement. *Capsella* is a fairly mild member

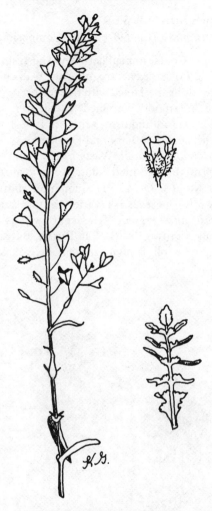

Fig. 68. SHEPHERD'S PURSE *(Capsella bursa-pastoris).*

of the mustard group and may be used in similar ways to Black Mustard. The leaves are used in salads or dried and added to soups and other dishes as a seasoning. The root may be dried and used as a substitute for ginger. Horned Larks and Lawrence Goldfinches eat quantities of the seed.

111. Common SUNFLOWER
Helianthus anuus (fig. 69) Family: Compositae

Description: Coarse annual herb with short, stiff hairs and simple leaves. Lower leaves are opposite, 6-15 cm long, and often somewhat heart-shaped with fine teeth along the edges. Upper leaves are smooth margined. The stems reach 3-20 dm in height. The yellow and brown flowers are 2-3.5 cm across and borne between February and October.

Distribution and Habitat: Waste places along roads up to 5000 ft (1500 m) throughout the state except in northern coast ranges and Mojave Desert. The common cultivated sunflower grown for its seeds is a variety of this same species.

Uses—Past and Present: The sunflower is native to America, but is grown for its oil in Russia, Bulgaria, India, and China. The oil is used in making soaps, candles,

Fig. 69. Common SUNFLOWER *(Helianthus annuus).* 1/2 X.

varnishes, and paints. It is also popular for cooking oil, being high in polyunsaturated fatty acids. The meal that remains after the oil is extracted is a high-protein supplement for poultry and livestock and is used in bird feeders. The hulls have been used for making a fuel log and for poultry litter. The seeds are nearly 55 percent protein and rich in vitamins and minerals especially vitamin B and calcium.

Indians of the Southwest gathered our native species and ground the seeds into flour. Some tribes cultivated the plant. The Canadian Ojibwas crushed the root between stones and applied the plant as a wet dressing for treating blisters. The Pimas used the inner stalk as a chewing gum or string for candles. They also made an extract of the leaves which they used to treat fevers. Sunflowers appear frequently in myths and decorations of the Southwestern Indians. The Hopis made yellow dye from the flowers and black and purple dyes from the seeds. These were used in basketry and cloth and body decoration. For collecting the seeds, most tribes would beat the flower heads when ripe, catching the seeds in baskets. These were then winnowed, parched, and ground into meal. Crushed seeds were sometimes boiled for their oil.

The pith from the stems of this versatile plant is employed as a mounting medium for microscope slides. The leaves have been dried and used as a substitute for tobacco in cigars, while in Germany, a fine writing paper is produced from the silky fiber obtained from the stalks. The buds of the sunflower are sometimes eaten like artichokes—boiled with butter and salt. Foragers can harvest the seeds by picking the dried flower heads before the seeds are released. Spread them out to dry in the sun or place them in a warm oven. When dry, beat, tramp, or shake the seeds loose and winnow out the trash. If a food grinder is available, select a setting that will just about let the seeds go through. The hulls will then be broken. Stir these into water for a half hour and the kernels will sink and the floating hulls may be skimmed off. Nuts can then be dried slowly in an oven or used wet in many popular recipes. They are nutritious and delicious when ground into flour or meal and added to baked products. Some foragers prefer to grind hull and all, for the hulls are quite thin.

Goldfinches are particularly fond of sunflower seeds, while game birds and songbirds also eagerly seek them out. Several species of *Helianthus* are on the rare and endangered plant list for California.

SUNFLOWER MUFFINS

3/4 cup sunflower
  seed meal
1 cup whole wheat flour
1 Tbsp baking powder
3/4 tsp salt

1/2 tsp mace (optional)
3 Tbsp brown sugar
1 egg, beaten
3/4 cup sour cream
  or buttermilk
1/4 cup milk

Preheat oven to 375° (quick mod.). Mix all dry ingredients in one bowl and all wet ingredients in another bowl. Combine the wet with the dry until just blended. Bake in greased muffin tins about 15 minutes. Makes 1 dozen.

SUNFLOWER SEED PANCAKES

2 pkg active dry yeast
1 1/2 cups warm yoghurt
  or buttermilk
1 Tbsp brown sugar
  or honey
2 Tbsp oil or melted butter

1/4 cup sunflower
  seed meal
1 1/2 cups whole wheat
  flour
3/4 tsp salt

Stir yeast into warm yoghurt or buttermilk. Add remaining ingredients and stir until blended. Drop onto ungreased hot griddle and turn when bubbles appear. Cook slowly. Delicious with honey or Strawberry jelly. Serves 4.

112. Milk THISTLE
Silybum marianum (pl. 7d) Family: Compositae

Description: Stout annual herb with large prickly lobed leaves that have conspicuous white mottling on the upper surface. Plant is 1-2 m tall, leaves 3-7 dm wide with clasping bases bearing many yellow prickles. Flower is purple with numerous spiny protruberances at its base. The head is 2.5-5 cm across and is produced from May to July.

Distribution and Habitat: A common weed in waste places, pastures, and fields throughout most of the state.

Uses—Past and Present: The thistles, which include several genera, are related to artichokes and may be treated in the same manner. The genera in this group include *Cirsium, Cynara, Centaurea,* and their many species. The Cardoon, *Cynara cardunculus,* is one of the largest and most delicious of the thistles. It is the plant from which the cultivated artichoke was derived. One species of *Cirsium* was a life-saving plant to Truman Everts who was lost for a month in the region of Yellowstone Park in 1870. He subsisted on the roots. In general, I try all thistles that are large enough to gather. They are boiled until tender and eaten like the artichoke. Both the young flowering stalks and roots of *Cirsium vulgare,* the Bull Thistle, have been found delicious when boiled 20 minutes and seasoned with salt and butter. The Milk Thistle is widespread and delicious. The young leaves, stalks, and roots can be soaked overnight in salted water and then cooked and eaten. The thistle heads are best when picked just at the height of flowering, boiled, and the heart (actually the receptacle) dissected out. A substitute for rennet used in coagulating milk for cheesemaking, and so on, can be obtained from the dried flowers of some thistles.

THISTLE HEARTS AU GRATIN

2 cups thistle hearts	1/2 tsp salt
1/4 cup butter	1/2 tsp white pepper
or margarine	1/4 tsp dry mustard
1/3 cup flour	1/4 cup toasted slivered
1 1/2 cups milk	almonds
6 oz. Gruyére cheese, shredded	

Preheat oven to 450° (hot). Snip thistle heads from plants with scissors, boil until tender, and remove hearts, reserving liquid. (Tenderness is judged by the ease with which a fork pierces the bottom.) Melt butter or margarine in saucepan and stir in flour, until smooth and bubbly. Add milk and 1/2 cup water in which the thistles were cooked and stir until thickened. Add cheese, salt, pepper, and mustard and cook, stirring, until cheese melts and sauce is smooth. Arrange thistle hearts in a casserole dish, pour sauce over, and sprinkle with almonds. Bake 15 minutes. Serves 4.

CHICKEN JERUSALEM

1 3-lb fryer, cut up	2 cups (or less) thistle
1/4 cup milk	hearts, cooked
1/3 cup flour	2 cups vanilla yoghurt,
6 Tbsp butter	with 1/2 tsp salt,
or margarine	with a little nutmeg
salt and pepper	paprika

Preheat oven to 350° (mod.). Dip chicken pieces into milk, then roll in flour. Brown on all sides in hot butter in skillet. Put in 3-qt casserole dish and sprinkle with salt and pepper. Spread thistle hearts on top of chicken and season with salt and pepper. Cover and bake for 45 minutes to 1 hour. Add yoghurt or sour cream, cover, and bake for 10 minutes longer. Sprinkle with paprika. Serves 4.

THISTLE STALKS AND APPLES

1 lb young thistle stalks	1/4 cup butter
1 cup boiling water	or margarine
1 tsp salt	2 medium green apples,
	peeled, cored, and
	sliced thin

Peel and cut thistle stalks into 2- to 3-in pieces. Boil in salted water until nearly tender (about 20 minutes). Drain and keep warm. In a skillet, over moderate heat, melt butter or margarine and sauté apples until tender but not mushy (about 10 minutes). Add stalks to apples and mix well. Serves 4-6.

113. Russian THISTLE
(also Tumbleweed, Prickly Saltwort) *Salsola iberica* formerly *S. kali* (fig. 70) Family: Chenopodiaceae

Description: Annual, many-branched herb 3-10 dm high. Forms a round, bushy clump, so familiar when it is blown by the wind up against fences. Leaves are very small, round in cross-section, and linear so the plant appears leafless. Flowers are small, greenish, and often tinged with red. It is most abundant in summer.

Distribution and Habitat: Common in cultivated fields and waste places as well as most plant communities below the Pine belt throughout the state.

Uses—Past and Present: Indians discovered early that this unsavory looking prickled bush from Eurasia is a savory vegetable when the early shoots are picked. Boiled and eaten alone with butter or chopped raw into salads, the tender shoots are relished by housewives in the Midwest, but little

Fig. 70. Russian THISTLE *(Salsola iberica).* 1/2 X.

used elsewhere. In Europe, the ashes of this plant were once used in the production of a carbonate of soda known as Barilla. Tumbleweed has an excellent flavor and is a very good cooked vegetable that is welcome on my table all year. The tenderest branch tips can be clipped from older plants that have had access to water, leaving the plant to be used most of the year. It is related to spinach and may be made into soups and added to other dishes as one would Celery. The seeds provide food for numerous birds and rodents, while Mule Deer feed on the young plants.

CREAMED TUMBLEWEED

1/4 cup butter or bacon fat
2 qt tender tumbleweed
 stems and leaves, cut up
1 tsp salt
1/2 tsp basil leaves

1/2 cup boiling water
1/2 cup light cream
 or half-and-half
1/8 tsp pepper

Heat butter or bacon fat in a skillet or heavy saucepan, add Tumbleweed, salt, basil, and boiling water. Cover and cook until tender. Add cream and pepper and serve at once. Serves 4-6.

114. Sow THISTLE
Sonchus oleraceus (cover photo, upper right)
Family: Compositae

Description: Leafy, stemmed annual 0.5-2.5 m tall. Lower leaves have a petiole and are 1-2 dm long with a lyre shape and teeth on the lobes. The lobe at the tip of the leaf is triangular. Upper leaves are reduced and clasp the stem. Flowers are yellow and resemble those of Dandelion. Flowers most of the year.

Distribution and Habitat: This is a common weed related to Lettuce. It is found in waste and moist places throughout the state.

Uses—Past and Present: Sow Thistle was mentioned by Dioscorides as an edible herb. The natives of New Zealand, North Africa, Germany, and Arabia seek it for food. After its introduction to North America, the Indians soon learned to relish the young plants. As it matures, the bitterness increases and it must be boiled in two changes of water. A gum obtained from the bitter, milky sap was once used to treat opium addiction. Much of this gum is eliminated if the young stalks are peeled of their outer skin. Boiled like asparagus, they make a delicious vegetable. The greens have some of the bitterness of endive or Dandelion, but mixed with other vegetables they are wholesome and delicious. The young leaves steamed and flavored with lemon and butter are quite pleasant. These leaves make excellent soups or souffles. The common name is believed to be derived from the fact that pigs eagerly consume the plant.

SOW THISTLE SOUFFLE

1 Tbsp butter or margarine	1 tsp salt
1 Tbsp lemon juice	dash of pepper
1 Tbsp flour	1 cup Sow Thistle leaves, cooked and chopped
1/2 cup milk	4 eggs, separated
	1/2 cup shredded Swiss or Monterey jack cheese

Preheat oven to 350° (mod.). Combine all ingredients except egg whites and Sow Thistle leaves. Heat gently so as not to curdle the eggs. Beat egg whites until stiff. Fold all ingredients together gently and pour into a deep, greased casserole dish. Bake for 30 minutes or until puffed and light brown. Serve with a cheese or lemon white sauce. Serves 6.

SONCHUS FRITTATA

1 qt Sow Thistle leaves, torn into pieces	2 cups diced cooked turkey
4 slices bacon, diced	8 eggs
3 Tbsp olive oil	1 tsp paprika
1 onion, minced	1/2 tsp poultry seasoning
	salt and pepper

Boil Sow Thistle for 5 minutes and drain. Chop fine. Brown bacon in oil and add onion and turkey. Cook until onion is tender but not brown. Beat eggs slightly with paprika, poultry seasoning, and salt and pepper to taste. Pour into skillet. Add Sow Thistle and cook until creamy, stirring constantly. Serve with mushrooms or tomato sauce if desired. Serves 4-6.

115. YERBA SANTA
(also Mountain Balm) *Eriodictyon californicum*
(fig. 71) Family: Hydrophyllaceae

Description: Aromatic shrub, evergreen, with shredding bark and weedy growth. Leaves alternate, somewhat leathery, 0.5-1.5 dm long and often toothed. They are sticky above and light-colored below. Flowers occur in terminal branched scorpioid cymes, lavender to white, tubular or horn-shaped, 9-15 mm long, and have 5 small lobes and 5 stamens. Blooms from May to July.

Distribution and Habitat: Below 5500 ft (1700 m) on dry slopes in Chaparral, Foothill Woodland, Yellow Pine Forest, Mixed Evergreen Forest, Oak Woodland, and Redwood Forest. This species is found mostly in northern California, but a similar species that has been used in the same way is *E. trichocalyx,* which occurs in southern California.

Uses—Past and Present: Indians and early settlers used Yerba Santa leaves as a remedy for colds, grippe, and asthma. It was either smoked or made into a tea by the Miwoks, Pomos, and Yukis of northern California. At times, both leaves and flowers were steeped for tea and drunk for the above ailments as well as stomachache and rheumatism. They were also warmed and used as a poultice on aching or sore areas. Mashed leaves were often applied to cuts, wounds, abrasions, and fractured bones to keep swelling down and aid in mending as well as relieve pain. Yerba Santa was adopted for many of these uses by the Spanish missionaries and entered into the U.S. pharmacopoeia in 1894.

To make tea, tear up 2-3 fresh or dried leaves, and pour boiling water over them. Cover and let steep. Chewing the fresh leaves produces a refreshing taste in the mouth. The mild bitterness first experienced is soon dissipated. *E. angustifolium* of San Bernardino Co. and *E. traskiae* of the

Fig. 71. YERBA SANTA
(Eriodictyon californicum). 2/3 X.

Channel Islands are listed on the California Native Plant Society's rare and endangered plant list and should be respected.

PLANTS TREATED ELSEWHERE WHICH
MAY ALSO OCCUR IN URBAN AND CULTIVATED
AREAS
(Refer to General Index for page numbers)

Beavertail Cactus
Blackberry
Cattail
Celery
Currant
Elderberry
Fennel

Lemonade Berry
Wild Mint
Oak
Peppermint
Pine
Rose
See also Ornamentals

ORNAMENTALS

116. CARNATION
(also Cottage Pink) *Dianthus plumarius* (fig. 72)
Family: Caryophyllaceae

Description: Low, tufted, hairless plant with small, carnationlike flowers that are scented in various ways. Similar to Sweet William *(D. barbatus).* It is 1.5-3 dm high, basal leaves 2.5-10 cm long. Leaves are narrow, linear, and acute, with a prominent midrib. Flowers rose or pink with clawed petals. These small, fragrant Carnation flowers are variable. Other flowers in the carnation group may be used in a similar manner. Blooms appear in late spring and early summer.

Fig. 72. CARNATION *(Dianthus plumarius).* 3/4 X.

Distribution and Habitat: Dianthus of many types thrive in full sun with fast draining soil. The Cottage Pinks need a fairly rich soil.

Uses— Past and Present: The Gillyflower of Shakespeare's time and Sops-in-Wine of Chaucer's were what we know as Carnation. For centuries, the Pinks have been prized for their clovelike flavor and fragrance. They were popular especially in seventeenth- and eighteenth-century England and France, where butters, cordials, syrups, vinegars, and other products were made with them.

CARNATION MARMALADE

1/2 lb fresh Dianthus flowers	1 cup sugar
	1 cup water

Trim base of petals to remove bitter white part. Crush petals. Put sugar and water in a saucepan and boil down to a syrup. Add crushed petals and simmer until pulpy. Stir well, pour into jars, and seal. Marmalade has the consistency of honey.

CARNATION MEAT SAUCE

2 cups Dianthus petals, firmly packed	1/2 tsp ground cloves
1 qt wine vinegar	1/4 tsp cinnamon
1 cup sugar	1/4 tsp nutmeg

Trim bitter base from petals. Place all ingredients in a 2-qt saucepan. Bring to a boil, then simmer, stirring, for 1 hour. Strain. Cool and store in refrigerator and use cold with meats such as ham or venison.

117. CAROB

(also Algaroba, Locust Bean, St. John's Bread)
Ceratonia siliqua (fig. 73) Family: Leguminosae

Description: Evergreen tree with thick crown. Leaves are compound with 2-3 pairs of oval, shiny leaflets, often with a notch at their tip. Flowers are small, borne on the old wood.

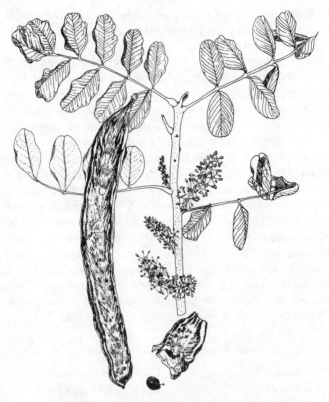

Fig. 73. CAROB *(Ceratonia siliqua).* 1/2 X.

No petals and 5 stamens. Pod is like a thick bean pod, 10-26 cm long, leathery, and filled with sweet pulp between seeds. Color of pod is dark red-brown. Pods are produced in fall but often hang on the trees until much later in the year.

Distribution and Habitat: The tree is native to the eastern Mediterranean and is now widely cultivated in this state as an ornamental.

Uses—Past and Present: The pods have long been used for both human and cattle food. Evidence of its ancient uses exists in preserved specimens in the museum at Naples where remains were exhumed from a house in Pompeii. The dried pods contain 50 percent sugar and are sold in some cities like candy. The United States imported over 1,250,000 pounds in

1935 for use in flavoring dog biscuit and chewing tobacco. It is often sold now as a substitute for chocolate in many products.

In preparing numerous delicious products from the Carob pods, the hard seeds must first be removed. This is best accomplished if the pods are picked before they harden. Otherwise, they may be soaked or simmered until soft enough for the seeds to be extracted. To make Carob meal, pare off the stem end, then place the entire pod in water and grind it to bits in a blender. Simmer the meal for about 10 minutes and strain. Discard the water and spread the meal thinly on a cookie sheet or screen to dry in a warm place or pilot-lit oven. When thoroughly dry, the meal can be ground in a hand gristmill or flour mill to a fine powder. A fair flour can also be made using the blender as described in the Notes on Plant Preparation Techniques. The water extraction described above is not necessary, but improves the product. For a sweeter meal, add 1 part sugar to four parts Carob flour. This powder can be added to milk and heated for a cocoa substitute or used in recipes calling for chocolate.

SOUR CREAM CAROB POUNDCAKE

1 cup butter
  or margarine
3 cups brown sugar
1 1/2 cups Carob meal
6 eggs
2 cups flour

1/2 tsp baking soda
1 cup sour cream
1 tsp rum
  or vanilla flavoring
1 cup chopped nuts
  (optional)

Preheat oven to 325° (slow mod.). Cream together butter or margarine and sugar. Add Carob meal and eggs, then blend in remaining ingredients until smooth. Bake for 1 1/2 hours in a greased loaf pan.

CAROB-HONEY BROWNIES

1/3 cup butter
  or margarine
3/4 cup honey

3/4 cup finely ground
  Carob meal
2 eggs, beaten

3/4 cup flour
1/2 tsp baking powder
1/4 tsp salt

1 tsp vanilla
1/2 cup chopped nuts
(optional)

Preheat oven to 350° (mod.). Melt butter or margarine and mix in honey and Carob meal. Beat together eggs and remaining ingredients until smooth. Bake in greased 8x8-in pan for 35 minutes.

For a quick glaze for either of the above recipes, mix sweetened Carob flour with sour cream to taste or sprinkle with powdered sugar.

NO-BAKE HEALTH FUDGE

1 cup honey
1 cup peanut butter
1 cup Carob flour
1 cup shelled
 sunflower seeds

1/2 cup toasted
 sesame seeds
1/2 cup flaked coconut
1/2 cup chopped walnuts
 (or other nuts)
1/2 cup raisins

In a large saucepan, heat honey and peanut butter, stirring constantly. When smooth, remove from heat and stir in Carob flour. Mix well and add remaining ingredients. Press into a buttered 8x8x2-in pan. Chill several hours or overnight. Cut into 1-in squares. Store in refrigerator. Makes 2 1/2 lbs of fudge.

118. CHRYSANTHEMUM

Chrysanthemum morifolium (fig. 74) Family: Compositae

Description: This is the "mum" of florists. Stout branching 0.5-1.5 m tall, herbage gray-pubescent. Leaves large, thick, heavy, and lobed. Heads of flowers clustered. Flowers are yellow, pink, lilac, red, white, with numerous forms. Blooms in autumn.

Distribution and Habitat: Chrysanthemum does best when planted in well-drained garden soil improved by organic matter and fertilizer away from trees or hedges with invasive

Fig. 74. CHRYSANTHEMUM *(Chrysanthemum morifolium)*. 1/2 X.

roots. It will take full sun unless in a very hot area and needs only moderate watering.

Uses—Past and Present: Cultivated in China as early as 500 B.C., Chrysanthemum was prized for boths its beauty and its food. Both leaves and flowers were eaten. Pick flowers in morning and pull off petals, removing the base of the petal. For use in salads, they should be dipped in boiling water and sweetened slightly. Add to fruit or green salads, chicken dishes, or soups. Do not overcook.

MUM SALAD

1/2 cup Chrysanthemum
  petals
1 1/2 cups sliced
  fresh mushrooms

1 small head lettuce,
  torn into pieces
1 hard-cooked egg,
  sliced

Toss ingredients together and serve with french dressing or oil and vinegar.

MUM CHOWDER

petals from 1 doz medium-sized Chrysanthemum flowers	4 slices bacon, chopped 1 medium onion, chopped 2 Tbsp butter
juice of 1 lemon	or margarine
1 tsp salt	1 qt milk
1 potato, peeled and chopped	1 cup chopped clams salt and pepper to taste

Tear off base of mum petals and toss into a bowl of cold water to which 1 tsp salt and the juice of 1 lemon have been added. Rinse, drain, and pat dry. Boil the potato until tender. Add bacon and boil for 2 more minutes. Drain. Sauté onion in butter until tender. Combine all ingredients except petals and simmer, covered, for 10 minutes. If a stronger chowder is desired, add 1 cup clam juice in place of 1 cup milk. Just before serving, add the flower petals. Serves 4.

119. EUGENIA
   (also Australian Brush Cherry) *Syzygium paniculatum* formerly *Eugenia paniculata* (pl. 7e)
Family: Myrtaceae

Description: Evergreen tree to 12 m or more, often grown as a hedge. Leaves opposite, oblong, 3.5-7.5 cm long, glossy, and reddish when young. Flowers white, 1.5-2.5 cm across in axillary bunches or on ends of short branches showing above foliage. Fruit an oval rose-purple berry about 2 cm in diameter. The flowering time is variable.

Distribution and Habitat: Eugenia grows best in well-drained soil with ample water. Sheltered spots or partial shade is preferred. When pruned into a hedge, the flowering and fruiting is reduced.

Uses—Past and Present: The clove is the bud of a *Syzygium*. This fruit is often used for nibbling or to make jelly.

EUGENIA JELLY

4 lb ripe Eugenia fruits   1 pkg powdered pectin
1 cup water   4 1/2 cups sugar
1 cup lemon juice

Wash and crush fruit and add water and 1/2 cup lemon juice. Cover and simmer for 15 minutes. Run through a jelly bag or moistened cheesecloth. (You will need 3 cups of juice.) Add 1/2 cup lemon juice and pectin to extract and stir well. Bring to a boil and add sugar, stirring constantly at a rolling boil. Boil exactly 2 minutes, skim, pour into sterilized jars, and seal with paraffin. If a sauce is desired, to each cup of juice, add sugar and lemon mixture to taste and stir in 1 Tbsp cornstarch dissolved in a little water. Simmer and stir until clear. This may be refrigerated and used like jelly.

120. Lemon-Scented GERANIUM
Pelargonium crispum (fig. 75) Family: Geraniaceae

Description: Woody, 6-9 dm high, pubescent, grayish, and leafy shrub. Leaves alternate, small, usually not exceeding 2.5 cm long. Leaves have teeth on edges, otherwise resemble small typical geranium leaves. Flowers 2-2.5 cm long, pink or rose with darker markings. This and numerous other species of *Pelargonium* are grown for their scented foliage.

Distribution and Habitat: Geranium grows well in pots and flower beds, requiring full sun, fast-draining soil, and little water.

Uses—Past and Present: In Africa, numerous species of *Pelargonium* are eaten. Colonists used the various scented geraniums in dry bunches for scenting closets, potpourris, and finger bowls. A number of them were also eaten. They may be used to line cake pans for getting the essence into the batter. Flavor jellies by adding leaves to the jar when the hot jelly is poured in. Crushed leaves may be used for flavoring soups, poultry, fish, sauces, and fruit dishes. To make geranium-spiced vinegar, place 1 ounce of fresh geranium leaves in a pint bottle and cover with white vinegar. Cover tightly and set in the sun for 2 weeks.

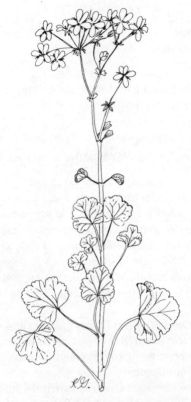

Fig. 75. Lemon-Scented GERANIUM *(Pelargonium crispum).* 1/2 X.

GERANIUM HERB SAUCE

1/4 cup butter	1/2 tsp salt
or margarine	8 Geranium leaves
1/4 cup flour	2 tsp parsley flakes
2 cups milk	(or 2 Tbsp chopped
	fresh parsley)

Melt butter over low heat in saucepan and stir in flour to make a smooth paste. Add milk all at once and, stirring constantly, add leaves, salt and parsley. Cook until thickened (if too thick, stir in a little more milk), remove leaves, and serve with peas, celery, carrots, corn, or cauliflower. This sauce may use any of the scented geraniums.

GERANIUM BUTTER

3/4 cup softened   2 tsp minced geraniums
  butter or margarine   leaves
1 1/2 tsp salt

Combine all ingredients and chill for several hours before using on vegetables or breads.

121. Pineapple GUAVA
Feijoa sellowiana (fig. 76) Family: Myrtaceae

Description: Attractive shrub to 5.5 m high. Leaves opposite, oval 5-7.5 cm long, glossy green above and silvery gray with short white hairs beneath. Flowers are solitary with 4 petals and many long exserted stamens. Petals are white inside; purplish-red outside. Stamens and style are dark red. Fruit is round, 2.5-7.5 cm long, dull green or tinged with red, and with a whitish bloom. Whitish flesh surrounds a jellylike pulp. Blooms May or June, with fruit ripening 4-5 months later.

Distribution and Habitat: Plants grow well in valley heat but fruits are better flavored in cooler coastal areas. It can take any amount of pruning or training and is considered one of the hardiest of subtropical fruits.

Uses—Past and Present: Originating from South America, the Pineapple Guava is favored both for its beautiful flowers and leaves and for its sweet fruits. The sweet flower petals can be used raw in salads, while the fruits can be used for jam or jelly or baked in pies or cakes like pineapple. The plant is bird pollinated.

PINEAPPLE GUAVA PIE

2 cups ripe Guava pulp   8 oz cream cheese
1/3 cup water   2 Tbsp heavy cream
2 Tbsp lemon juice   2 Tbsp powdered sugar
2 Tbsp honey   1 cookie or graham
                                  cracker crust

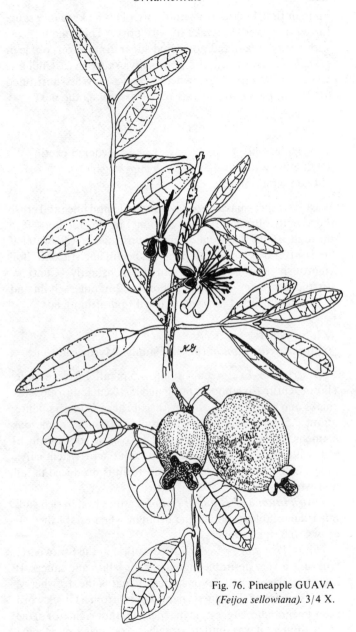

Fig. 76. Pineapple GUAVA
(Feijoa sellowiana). 3/4 X.

Simmer first 4 ingredients until soft. (This makes a nice sauce for serving over shortcake or with pork.) Thin cream cheese with heavy cream and powdered sugar. Spread on cookie or graham cracker crust and pour guava sauce over. Chill and serve. Scrumptious! Strawberry Guava may be substituted for Pineapple Guava in this recipe as well as the next.

PINEAPPLE GUAVA JELLY

3 lb ripe guavas	1 pkg powdered pectin
3 1/2 cups water	5 cups sugar
1/4 cup lemon juice	

Wash and slice guavas. Add water and lemon juice and crush thoroughly. Bring to boiling point but do not boil. Run through a jelly bag, collecting 4 cups juice. (Add water if necessary.) In a large saucepan, add pectin and stir well. Boil vigorously and add sugar, stirring constantly. Continue stirring and bring to a rolling boil for 2 minutes. Skim and pour into sterilized jars and seal with paraffin or lids.

122. Strawberry GUAVA
Psidium cattleianum (pl. 7f) Family: Myrtaceae

Description: Evergreen shrub to 4.5 m tall. Leaves simple, elliptic, entire (or smooth) margined, 5-10 cm long, opposite, thick, and leathery. New growth is bronze. Flowers white, about 2.5 cm across with 4-5 distinct petals and numerous stamens. Fruit is fleshy and oval, 2.5-3.5 cm across, purplish-red, with white flesh. It is crowned with the persistent calyx lobes. One variety has yellow fruit. Fruit is produced in fall and winter.

Distribution and Habitat: Psidium grows best in rich soils but is adaptable, taking some drought when established. It grows well in containers.

Uses—Past and Present: Native to Brazil, the Strawberry Guava is a very desirable ingredient for jellies and sauces. It has a consistency and flavor similar to that of the strawberry, but it is creamier. Its generic name comes from a Greek word for pomegranate, suggesting that perhaps the originator of the name thought it was similar. Strawberry Guavas are welcome additions to fruit dishes of all sorts and make excellent jellies and jams if treated similar to peaches or other fruits. They are also tasty when chilled and eaten fresh. Strawberry Guavas can be used with recipes for Pineapple Guava.

STRAWBERRY GUAVA SYRUP

1 qt guavas	lemon and sugar to taste
1 1/2 cups water	cornstarch
1 cup honey or sugar	

Wash and mash guavas in a saucepan with water and honey or sugar. Add lemon and more sugar to taste and bring to a simmer for 20 minutes, until sauce has cleared somewhat. Strain through a tea strainer and discard pulp. To each cup of sauce, add 1 Tbsp cornstarch dissolved in a little water. Return to heat and simmer, stirring, until of desired consistency. Refrigerate. For variations, season with orange juice or your favorite spice. This is excellent on pancakes or ice cream.

123. Day LILY
   (also Golden Summer Day Lily) *Hemerocallis aurantiaca* (pl. 8a) Family: Liliaceae

Description: Large orange or yellow-flowered Lily with 5-8 leaves per stem. Leaves as long as 6-9 dm, 2.5 cm or more broad, and keeled. Has 6-8 flowers, fragrant, 7.5-10 cm long. Flowers are present for only 1 day. Double-flowered varieties are of equal quality. They bloom from May to September.

Distribution and Habitat: There are numerous varieties and species of day lily, ranging from yellow to red; all are edible. This species and others seem to have originated in Japan, and some have escaped from cultivation in disturbed areas. Another popular orange day lily is *H. fulva* which is deciduous and larger. It is persistent and pest-free, while adapting to almost any kind of soil. It grows in sun or partial shade.

Uses—Past and Present: In the Orient, day lilies are prized for food, especially with pork and soy sauce. Roots, buds, and flowers are all eaten. The roots taste like a combination of beans and corn to some people—radish and carrot to others. They may be eaten raw or cooked and are especially good when mixed with other vegetables in a salad. Buds may be cooked and served with herbed butter or added to soups and stews. The Chinese and Japanese consider the bean-flavored buds an important food. I enjoy them raw as well as cooked. Day Lily flowers are frittered by dipping into a plain batter and frying in fat. Serve hot, sprinkled with powdered sugar. Withered flowers are also good for soups and stews. Flowers and buds may be dried by placing in a warm place on newspaper, then stored in clean jars. To rehydrate, merely sauté in butter or soak in water for a few minutes. The young flowering stalks may be eaten like asparagus.

LILY SKILLET

3/4 cup day lily buds (dried or fresh)	1/4 cup water
1 cup sliced pork or pork sausage	1 tsp salt
	2 tsp sugar
3 Tbsp butter or margarine	1 Tbsp sherry
1 1/2 cups sliced mushrooms	2 Tbsp soy sauce

If using dried buds, soak in 1 cup of water for 20 minutes prior to using. Sauté pork in butter or margarine until brown. (If using pork sausage, omit butter or margarine, sauté, and drain.) Add remaining ingredients and cook until all are tender (about 5 minutes). Serves 6.

LILY ROOT BAKE

1 cup milk	1 small onion, minced
1/2 cup mayonnaise	1 tsp parsley flakes
1 egg, beaten	1 cup dry bread crumbs (or leftover stuffing, crumbled)
2 cups firm lily roots, cleaned and sliced	
1 cup herb-seasoned bread stuffing	2 Tbsp butter or margarine

Preheat oven to 350° (mod.). Combine milk and mayonnaise and add egg, lily roots, stuffing, onion, and parsley. Pour into greased and floured 8-in baking dish. Toss bread crumbs with melted butter and sprinkle over lily mixture. Bake for 30 minutes. Serves 6.

BLOSSOM FRITTERS

1 dozen lily blossoms	1/2 cup bread crumbs
1 egg	1/2 cup grated Parmesan
1 tsp water	cheese
	1/2 cup butter or margarine

Wet blossoms in egg beaten with water. Roll in mixture of bread crumbs and Parmesan cheese. Deep fry in melted butter or margarine and serve warm. Serves 6.

124. Pot MARIGOLD
   (also Calendula) *Calendula officinalis* (pl. 8b)
   Family: Compositae

Description: Annual with orange and yellow blooms 6-11 cm across. Also comes in shades of apricot, cream, and so on. Plants 3-6 dm high. Leaves are long, narrow, round on ends, and slightly sticky. Flowers in fall and winter.

Distribution and Habitat: Good in most soils, ample or little water if drainage is fast. Takes full sun.

Uses—Past and Present: Calendulas date back to early Roman times and were cultivated in England before 1573. The flower was used for making wine and giving color and flavor to other drinks and a variety of dishes. It was reported that those who partook of marigolds would be more amorous, see fairies, or be induced to sleep. They add color and texture to drinks and foods. Crisp the petals in ice water and garnish egg salads with them, or add chopped petals to salads and vegetable dishes just prior to serving. Add to any sauce in which you are cooking chicken, then garnish with chopped petals. They may be substituted for saffron in coloring rice or butter. The African Marigold, which is in the genus *Tagetes,* may also be used but is not as desirable.

Tagetes is the marigold commonly used by organic gardeners to repel pests.

MARIGOLD-ORANGE PUDDING

1 1/2 cups stale	3 Tbsp finely chopped
bread pieces	marigold petals
3 cups milk	2 egg whites
2 egg yolks, beaten	juice of 1 orange
1/2 cup sugar	1/4 cup sugar

Preheat oven to 375° (quick mod.). Soak bread, torn up, in milk. When bread has taken up most of the milk, add egg yolks, 1/2 cup sugar, and marigold petals. Bake in dish set in shallow tray of water for 45 minutes to 1 hour, or until custard is set. Beat egg whites until they peak, gradually adding orange juice and 1/4 cup sugar. Pile the meringue on top of the custard and brown quickly in the oven, just long enough to make stiff. Serves 6-8.

125. NASTURTIUM

Tropaeolum majus (fig. 77) Family: Tropaeol-aceae

Description: Perennial grown as an annual, climbing or spreading. Leaves with long petioles, orbicular in shape, 2-7 dm across with about 9 main nerves radiating from the center where the petiole is attached. Flowers yellow, red, white, and orange, 2.5-6 cm across with a 2.5 cm or longer spur. There are 8 stamens. Blooms in Spring.

Distribution and Habitat: Well drained soils in sun suit this common flower, but it grows best in sandy soil. Needs very little care.

Uses—Past and Present: Originally from Peru, the flowers and young leaves were grown for their attractiveness. Other species were grown for their berries and tubers in South America. Nasturtium was used in the vegetable gardens of England as early as 1726 and in Europe in 1574. Flowers and leaves are mixed into salads and the seed pods are used for pickling as a substitute for capers. They have a peppery flavor similar to the mustards. Flowers may be filled with cream

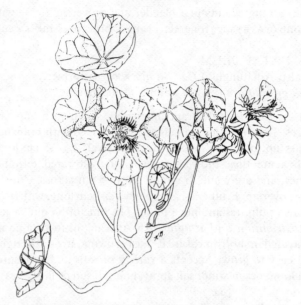

Fig. 77. NASTURTIUM *(Tropaeolum majus).* 1/2 X.

cheese and pineapple mixture for appetizers. Mix chopped leaves into a bean salad or coleslaw. Stuff blossoms with tuna fish salad.

PICKLED NASTURTIUM PODS

2 cups Nasturtium pods
2 cups white vinegar
10 peppercorns
1/2 cup minced onion
1 tsp salt
1/8 tsp nutmeg

Soak pods in salted water for 4 days and drain. Boil remaining ingredients for 10 minutes. Cool, strain, and pour over pods. Cover and let stand in refrigerator for 1 week.

NASTURTIUM DIP

2 egg yolks
1/4 tsp salt
1/2 tsp dry mustard
2 Tbsp lemon juice
1/2 cup oil
1 cup coarsely chopped
   Nasturtium leaves
1/4 cup chopped parsley
2 tsp Worcestershire sauce

Place all ingredients in a blender and pulverize. Serve with shrimp or as a sauce for green beans or broccoli. Makes 1 cup.

126. NATAL PLUM
(also Amatungula) *Carissa macrocarpa* formerly
C. grandiflora (pl. 8c) Family: Apocynaceae

Description: Stiff, branching, spiny shrub with opposite leaves. Reaches 5.5 m in height and is armed with branched spines up to 3.5 cm long. Leaves are oval, 2.5-7.5 cm long, with acute tips. Flowers occur in few-flowered terminal cymes, and are white, fragrant, about 5 cm across. Corolla lobes overlap. Fruit is a scarlet berry 2.5-5 cm long, with firm, reddish pulp, resembling a plum. The time of bloom varies.

Distribution and Habitat: An African shrub commonly cultivated in subtropical areas. Fast growing, strong, upright, and easy to grow. Accepts a variety of soils and exposures including ocean wind, salt spray, and full sun or fairly heavy shade.

Uses—Past and Present: Fruits vary in sweetness, but generally are rather like a sweet cranberry and make good sauce. They are eaten raw, but are more often used for jelly.

NATAL PLUM JELLY

4 lb ripe Natal Plums   1 pkg powdered pectin
1 cup water   6 cups sugar
1/4 cup lemon juice

Wash and crush Natal Plums. Add water and lemon juice and simmer, covered, for 15 minutes. Run through a jelly bag. You will need 4 cups of juice; add water if necessary. Add pectin and stir well. Bring to a rolling boil and add sugar stirring constantly. Bring back to a full boil and boil hard for 2 minutes, skimming if necessary. Pour into sterilized jars and seal with paraffin or double lids. If you prefer not to add pectin, the sauce may be thickened with cornstarch as in the Pineapple Guava recipe. It may then be used as a sauce, a syrup, or a topping.

127. PYRACANTHA
(also Firethorn) *Pyracantha coccinea* (pl. 8d)
Family: Rosaceae

Description: Evergreen shrub to 2 m. Leaves narrow, elliptic, 2-5 cm long, finely toothed. Flowers are white, 1 cm across, in a many-flowered cluster. Fruit is bright red, round, and about 5 mm in diameter. Flowers March through April, but fruit is available almost all year around. There are many varieties of this shrub.

Distribution and Habitat: Numerous species and varieties of *Pyracantha* are grown and can be used similarly. Plants do best in full sun and in partially dry soil. They are often clipped into hedges and topiary shapes. They are native to southern Europe and Asia Minor.

Uses—Past and Present: Pyracantha berries are a favorite for jelly. It is not necessary to remove all the small stems. The fruit is added to some citrus juices to heighten their flavor. Sauce made of the sweetened extract of the simmered berries complements meats, poultry, and some vegetables. Try poaching chicken in the jelly.

SPICED PYRACANTHA MARMALADE OR SAUCE

2 cups Pyracantha
  berries
1 cup water
dash of salt
grated peel, juice, and pulp
  of 1 orange

1 Tbsp lemon juice
1/2 tsp allspice
1/2 cup sugar
3 Tbsp cornstarch

Simmer berries, covered, in salted water for 10 minutes and strain. For each cup of juice, add the grated peel, juice, and pulp of 1 orange plus lemon juice. Add allspice and sugar to taste. Simmer 30 minutes until orange peel is tender. Add cornstarch dissolved in a little cold water, and simmer until thickened. Store in refrigerator. This is an excellent accompaniment to meats and sandwiches. For firmer jelly, triple this recipe and add 1 box of pectin instead of the cornstarch. Following the directions for making jam on the pectin box, boil hard for 4 minutes after adding the pectin, then put into jars and seal.

PLANTS TREATED ELSEWHERE WHICH MAY
ALSO OCCUR AS ORNAMENTALS
(Refer to General Index for page numbers.)

Agave	Elderberry	Oak
Alfalfa	Fennel	Onion
Barley	Mountain Grape	Fan Palm
Beet	Wild Grape	Peppermint
Barrel Cactus	Iceplant	Pine
Beavertail Cactus	Juniper	Rose
Cattail	Lemonade Berry	New Zealand
Holly-leaved	Live Forever	Spinach
Cherry	Manzanita	Watercress
Chicory	Wild Mint	Yucca

OBTAINING FURTHER ASSISTANCE

It is imperative that plants be correctly identified before they are sampled. Some information can be found under the common name of a plant, but much more may be obtained if you know the Latin names. This book gives the Latin name of each plant discussed. Once this information is known, the literature about that particular plant can give the remaining information needed to determine its edibility or preparation procedure.

Many weeds and ornamentals are known to nursery owners and landscapers. High school and college botany teachers will usually know the most common plants, and every university and college in the state has at least one botanist who can correctly identify most plants. For assistance from one of these people, take a specimen of the plant to him or her. If this is not convenient, press flower, fruit (if possible), and leaves of a small part of the plant in a book for a few days and then mail them, with a request for the Latin name, to the institution in care of the Botany Department. Be sure to always enclose a self-addressed return envelope. The botanist will mail back the Latin name which can then be checked with this or other texts as to its edibility.

Other methods of identifying plants are also available. I rarely pass a nature center, tourist center, or outdoor display area without stopping to see if a nature trail or other information exists. Nature trails exist in most of our major state parks and are invaluable sources of plant identification and lore. Many botanical gardens have labelled nature trails. A few examples of trails for specific habitats are: (Desert) Joshua Tree National Monument Headquarters in Twenty-nine Palms; Palm Springs Desert Museum in Palm Desert; Borrego Palm Canyon in Anza-Borrego State Park. (Chaparral and other communities) Santa Barbara Botanical Garden; Rancho Santa Ana Botanical Garden in Claremont.

Throughout the state there are botanical gardens where actual specimens can be compared to samples. Here is a list of institutions that can be helpful for written information or which have specimens of many plants:

Rancho Santa Ana Botanical Garden
1500 N. College Ave.
Claremont, CA 91711

University of California Botanical Garden
405 Hilgard Ave.
Los Angeles, CA 90024

Botanical Garden
University of California
Strawberry Canyon
Berkeley, CA 94720

California Native Plant Society
Suite D, 2380 Ellsworth St.
Berkeley, CA 94704
(This society has frequent field trips throughout
the state.)

California Rare Fruit Growers
Star Route, Box P
Bonsall, CA 92003

Santa Barbara Botanical Garden
1212 Mission Canyon Rd.
Santa Barbara, CA 93105

Descanso Gardens
Department of Arboreta
and Botanical Gardens
1418 Descanso Dr.
La Canada, CA 91011

Huntington Botanical Gardens
1151 Oxford Rd.
San Marino, CA 91108

Regional Parks Botanical Garden
  Tilden Regional Park
  Berkeley, CA 94708

South Coast Botanical Garden
  26701 Rolling Hills Rd.
  Palos Verdes Peninsula, CA 90274

Strybing Arboretum
  and Botanical Gardens
  Golden Gate Park
  9th and Lincoln Way
  San Francisco, CA 94122

University of California
  Botanical Garden
  Riverside, CA 92502

For the serious student, a course in field botany or ornamental horticulture is a good beginning. The study of ethnobotany or economic botany would also be helpful.

A good sourcebook listing many botanical gardens and horticultural societies is the *Directory of American Horticulture,* Revised Edition. For a copy, write to the American Horticultural Society, Inc., Mt. Vernon, VA 22121. The price is $5.00.

The following list of references will also serve as a working library for those who are interested in further research. Many of these books are written for a specific region of the country, however, and are not wholly suitable for California plants. For this reason, it is not recommended that the cookbooks among this list be purchased without prior knowledge of their contents.

Suggested References

Angier, Bradford, *Field Guide to Edible Wild Plants,* Stackpole Books, 1974.
———, *Free for the Eating,* Stackpole Books, 1966.
———, *More Free for the Eating Wild Foods,* Stackpole Books, 1969.

Baily, L. H., *Manual of Cultivated Plants,* Macmillan Co., 1949.

Balls, Edward K., *Early Uses of California Plants,* Univ. of California Press, 1962.

Barrett, S. A. and E. W. Gifford, *Miwok Material Culture,* Yosemite Natural History Association, 1933.

Barrows, David P., *The Ethno-Botany of the Coahuilla Indians of Southern California,* Malki Museum Press, Morongo Indian Reservation, 1967.

Bean, Lowell J. and Katherine Siva Saubel, *Temalpakh, Cahuilla Indian Knowledge and Usage of Plants,* Malki Museum Press, Morongo Indian Reservation, 1972.

Berglund, B. and C. Bolsby, *The Edible Wild,* Scribner's Sons, 1971.

Brackett, B. and M. Lash, *The Wild Gourmet,* D. R. Godine, 1975.

Bryan, John E. and Coralie Castle, *The Edible Ornamental Garden,* 101 Productions, 1974.

Burt, C. P. and F. G. Heyl, *Edible and Poisonous Plants of the Western States,* Lake Oswego, Oregon, 97034 (a deck of cards obtained directly from authors).

Castetter, Edward F., *Ethnobiological Studies in the American Southwest,* Univ. of New Mexico Bulletin, May 15, 1935.

Chapman, V. J., *Seaweeds and Their Uses,* Methuen & Co., 1970.

Christensen, Clyde M., *Common Edible Mushrooms,* Univ. of Minnesota Press, 1970.

Cronquist, Arthur, *Evolution & Classification of Lower Plants,* Houghton Mifflin, 1968.

Crowhurst, Adrienne, *The Weed Cookbook,* Lancer Books, 1972.

Dawson, E. Yale, *Seashore Plants of Southern California and Seashore Plants of Northern California,* Univ. of California Press, 1966.

DeLisle, Harold F., *Common Plants of the Southern California Mountains,* Naturegraph, 1961.

Enari, Leonid, *Ornamental Shrubs of California,* Ward Ritchie Press, 1962.

Gibbons, Euell, *Stalking the Blue-Eyed Scallop,* David McKay Co., 1964.

———*Stalking the Good Life,* David McKay Co., 1966.

———*Stalking the Healthful Herbs,* David McKay Co., 1966.

———*Stalking the Wild Asparagus,* David McKay Co., 1962.

Guberlet, M. L., *Seaweeds at Ebb Tide,* Univ. of Washington Press, 1956.

Hanle, Zack, *Cooking with Flowers,* Price/Stern/Sloan, 1971.

Harrington, H. D., *Edible Native Plants of the Rocky Mountains,* Univ. of New Mexico Press, 1967.

_____ and L. W. Durrell, *How to Identify Plants,* Sage Books, 1957.

Harris, Ben C., *Eat the Weeds,* Barre Publishers, 1972.

Hatfield, Audrey W., *How to Enjoy Your Weeds,* Collier Books, 1971.

Hedrick, U. P., ed., *Sturtevant's Edible Plants of the World,* Dover Publications, 1972.

Heizer, R. F. and M. A. Whipple, eds., *The California Indians,* Univ. of California Press, 1971.

Hill, Albert F., *Economic Botany,* McGraw-Hill Book Co., 1952.

Jaeger, Edmund C., *Desert Wild Flowers,* Stanford Univ. Press, 1940.

Kingsbury, John M., *Poisonous Plants of the United States and Canada,* Prentice-Hall, 1964.

Kirk, Donald R., *Wild Edible Plants of the Western United States,* Naturegraph, 1970.

Kroeber, A. L., *Handbook of California Indians,* California Book Co., 1962.

Lucas, Richard, *Nature's Medicines,* Award Books, 1966.

MacNicol, Mary, *Flower Cookery,* Collier Books, 1972.

Martin, A. C., H. S. Zim, and A. L. Nelson, *American Wildlife & Plants,* Dover Publications, 1951.

Medsger, Oliver P., *Edible Wild Plants,* Collier Books, 1972.

Munz, Philip A. and David D. Keck, *A California Flora,* Univ. of California Press, 1965, and Supplement, 1968.

Niehaus, Theodore F., *Sierra Wildflowers,* Univ. of California Press, 1974.

Niethammer, Carolyn, *American Indian Food and Lore,* Collier Books, 1974.

Ornduff, Robert, *Introduction to California Plant Life,* Univ. of California Press, 1974.

Orr, Robert T. and Dorothy Orr, *Mushrooms,* Univ. of California Press, 1968 and 1971.

Peterson, P. Victor, and P. Victor Peterson, Jr., *Native Trees of the Sierra Nevada,* Univ. of California Press, 1974.

Peterson, P. Victor, *Native Trees of Southern California,* Univ. of California Press, 1966.

Powell, W. Robert, *Inventory of Rare and Endangered Vascular Plants of California,* California Native Plant Society, 1974.

Rambsbottom, John, *Mushrooms and Toadstools,* Collins Press, 1954.

Raven, Peter H., *Native Shrubs of Southern California,* Univ. of California Press, 1966.

Saunders, Charles F., *Useful Wild Plants of the United States and Canada,* McBride & Co., 1920.

Schneider, Albert, *The Medicinal Plants of the California Indians,* Merck Report, 1906.

Smith, Alexander H., *Mushrooms in Their Natural Habitats,* Hafner Press, 1949.

Smith, Gilbert M., *Marine Algae of the Monterey Peninsula,* Stanford Univ. Press, 1964.

Steward, A. M. and Leon Kronoff, *Eating from the Wild,* Ballantine Books, 1975.

Sunset Magazine, ed., *Western Garden Book,* Lane Magazine & Book Co., 1967.

Sweet, Muriel, *Common Edible and Useful Plants of the West,* Naturegraph, 1962.

Tate, Joyce L., *Cactus Cook Book,* Cactus and Succulent Society of America, 1971.

Thomas, John Huntor, and Dennis R. Parnell, *Native Shrubs of the Sierra Nevada,* Univ. of California Press, 1974.

Thompson, Steven and Mary Thompson, *Wild Food Plants of the Sierra,* Dragtooth Press, 1972.

Weiner, Michael A., *Earth Medicine—Earth Foods,* Collier Books, 1972.

LIST OF PHYLOGENETIC RELATIONSHIPS
(See General Index for page numbers)

DIVISION Chlorophyta—Green Algae
    Ulva lactuca—Sea Lettuce

DIVISION Phaeophyta—Brown Algae
    Egregia laevigata—Feather Boa Kelp*
    Laminaria farlowii—Kombu*
    Macrocystis pyrifera—Giant Kelp*
    Nereocystis leutkeana—Sea Whip

DIVISION Rhodophyta—Red Algae
    Porphyra perforata—Laver

DIVISION Eucomycota—True Fungi
  CLASS Basidiomycete—Basidium Fungi, Club Fungi
    Lycoperdon perlatum—Gem-Studded Puffball
    Coprinus comatus—Shaggy Mane
  CLASS Ascomycete—Sac Fungi
    Morchella esculenta—Common Morel*

DIVISION Tracheophyta—Vascular Plants
  CLASS Filicineae—Ferns
   ORDER Filicales
    FAMILY Pteridaceae
     Pteridium aquilinum—Bracken Fern
  CLASS Gymnospermae—Gymnosperms
   ORDER Coniferales—Cone-bearing Plants
    FAMILY Pinaceae
     Pinus lambertiana—Sugar Pine*
     Pinus monophylla—One-leaf Pinyon Pine*
     Pseudotsuga menziesii—Douglas Fir*
    FAMILY Cupressaceae
     Juniperus californica—California Juniper
   ORDER Ephedrales
    FAMILY Ephedraceae
     Ephedra nevadensis—Nevada Mormon Tea*

*Additional species discussed in text.

CLASS Angiospermae—Flowering Plants
 SUBCLASS Dicotyledonae
  ORDER Magnoliales
   FAMILY Lauraceae
    Umbellularia californica—California Bay
  ORDER Ranunculales
   FAMILY Berberidaceae
    Mahonia pinnata—Mountain Grape*
  ORDER Urticales
   FAMILY Urticaceae
    Urtica dioica ssp. holosericea—Stinging Nettle*
  ORDER Juglandales
   FAMILY Juglandaceae
    Juglans californica—South California Black Walnut*
  ORDER Fagales
   FAMILY Fagaceae
    Chrysolepis sempervirens—Sierra Chinquapin*
    Quercus agrifolia—Coast Live Oak*
   FAMILY Betulaceae
    Corylus cornuta—California Hazelnut
  ORDER Caryophyllales
   FAMILY Cactaceae
    Ferocactus acanthodes—Barrel Cactus
    Opuntia basilaris—Beavertail Cactus*
   FAMILY Aizoaceae
    Gasoul crystallinum—Iceplant*
    Tetragonia tetragonioides—New Zealand Spinach
   FAMILY Caryophyllaceae
    Dianthus plumarius—Carnation*
    Stellaria media—Common Chickweed
   FAMILY Portulacaceae
    Claytonia perfoliata—Miner's Lettuce
    Portulaca oleracea—Purslane
   FAMILY Chenopodiaceae
    Atriplex hymenelytra—Desert Holly*
    Atriplex patula ssp. hastata—Halberd-leaved Saltbush*
    Atriplex semibaccata—Australian Saltbush*
    Beta vulgaris—Sugar Beet
    Chenopodium album—Lamb's Quarters*
    Salicornia subterminalis—Parish's Pickleweed*
    Salsola iberica—Russian Thistle
    Suaeda californica—Sea Blite*

FAMILY Amaranthaceae
  Amaranthus retroflexus—Green Amaranth*
ORDER Polygonales
  FAMILY Polygonaceae
    Oxyria digyna—Mountain Sorrel
    Rumex acetosella—Sheep Sorrel*
    Rumex hymenosepalus—Canaigre*
ORDER Malvales
  FAMILY Malvaceae
    Malva parviflora—Cheeseweed*
ORDER Violales
  FAMILY Violaceae
    Viola purpurea—Mountain Violet*
  FAMILY Fouquieriaceae
    Fouquieria splendens—Ocotillo
ORDER Capparales
  FAMILY Cruciferae
    Brassica nigra—Black Mustard*
    Cakile maritima—Sea Rocket*
    Capsella bursa-pastoris—Shepherd's Purse
    Caulanthus inflatus—Squaw Cabbage
    Descurainea pinnata—Tansy Mustard*
    Lepidium virginicum—Wild Peppergrass*
    Nasturtium officinale—Watercress
    Raphanus sativus—Wild Radish
ORDER Ericales
  FAMILY Ericaceae
    Arctostaphylos glauca—Bigberry Manzanita*
    Gaultheria shallon—Salal*
    Vaccinium nivictum—Sierra Bilberry*
    Vaccinium occidentale—Western Blueberry
    Vaccinium ovatum—California Huckleberry*
ORDER Rosales
  FAMILY Crassulaceae
    Dudleya saxosa—Live Forever*
  FAMILY Saxifragaceae
    Ribes cereum—Squaw Currant*
    Ribes roezlii—Sierra Gooseberry*
  FAMILY Rosaceae
    Amelanchier pallida—Western Service Berry*
    Fragaria vesca—Wood Strawberry*
    Prunus ilicifolia—Holly-leaved Cherry*

Prunus virginiana var. demissa—Western Chokecherry
Pyracantha coccinea—Pyracantha*
Rosa californica—Wild Rose*
Rubus leucodermis—White-Stemmed Raspberry
Rubus parviflorus—Thimbleberry
Rubus spectabilis—Salmonberry
Rubus ursinus—California Blackberry*
FAMILY Leguminosae
Acacia greggii—Catclaw
Ceratonia siliqua—Carob
Cercidium floridum—Palo Verde*
Medicago sativa—Alfalfa
Olneya tesota—Ironwood
Prosopis glandulosa var. torreyana—Mesquite*
ORDER Myrtales
FAMILY Myrtaceae
Feijoa sellowiana—Pineapple Guava
Psidium cattleianum—Strawberry Guava
Syzygium paniculatum—Eugenia*
FAMILY Onagraceae
Epilobium angustifolium—Fireweed*
ORDER Cornales
FAMILY Cornaceae
Cornus sessilis—Miner's Dogwood*
ORDER Euphorbiales
FAMILY Buxaceae
Simmondsia chinensis—Jojoba
ORDER Rhamnales
FAMILY Vitaceae
Vitis californica—Wild Grape*
ORDER Sapindales
FAMILY Anacardiaceae
Rhus integrifolia—Lemonade Berry*
Rhus trilobata—Squaw Bush
FAMILY Zygophyllaceae
Larrea divaricata—Creosote Bush
ORDER Geraniales
FAMILY Geraniaceae
Erodium cicutarium—Red stem Filaree*
Pelargonium crispum—Lemon-Scented Geranium*
FAMILY Tropaeolaceae
Tropaeolum majus—Nasturtium*

ORDER Umbellales
  FAMILY Umbelliferae
    Apium graveolens—Celery
    Conium maculatum—Poison Hemlock
    Foeniculum vulgare—Sweet Fennel
    Perideridia gairdneri—Gairdner's Squaw Root
ORDER Gentianales
  FAMILY Apocynaceae
    Carissa macrocarpa—Natal Plum
  FAMILY Asclepiadaceae
    Asclepias speciosa—Showy Milkweed*
ORDER Polemoniales
  FAMILY Solanaceae
    Lycium fremontii—Boxthorn*
  FAMILY Hydrophyllaceae
    Eriodictyon californicum—Yerba Santa*
ORDER Lamiales
  FAMILY Labiatae
    Marrubium vulgare—Horehound
    Mentha arvensis—Wild Mint
    Mentha piperita—Peppermint
    Mentha spicata—Spearmint
    Monardella odoratissima—Mountain Pennyroyal*
    Salvia columbariae—Chia*
ORDER Plantaginales
  FAMILY Plantaginaceae
    Plantago major—Common Plaintain
ORDER Scrophulariales
  FAMILY Acanthaceae
    Beloperone californica—Chuparosa
ORDER Dipsacales
  FAMILY Caprifoliaceae
    Sambucus mexicana—Desert Elderberry*
ORDER Asterales
  FAMILY Compositae
    Artemisia tridentata—Big Sagebrush*
    Calendula officinalis—Pot Marigold*
    Chrysanthemum morifolium—Chrysanthemum
    Cichorium intybus—Chicory
    Helianthus annuus—Common Sunflower*
    Silybum marianum—Milk Thistle*
    Sonchus oleraceus—Sow Thistle

Taraxacum officinale—Dandelion
SUBCLASS Monocotyledonae
 ORDER Alismatales
  FAMILY Alismataceae
   Sagittaria latifolia—Arrowhead
 ORDER Cyperales
  FAMILY Cyperaceae
   Cyperus esculentus—Nut-Grass*
   Scirpus robustus—Prairie Bulrush
  FAMILY Gramineae
   Avena fatua—Wild Oat*
   Hordeum vulgare—Common Barley
 ORDER Typhales
  FAMILY Typhaceae
   Typha latifolia—Common Cattail*
 ORDER Arecales
  FAMILY Palmae
   Washingtonia filifera—Fan Palm
 ORDER Liliales
  FAMILY Liliaceae
   Allium haematochiton—Wild Red Onion*
   Calochortus nuttallii—Nuttall's Mariposa Lily*
   Camassia leichtlinii—Leichtlin's Camas*
   Dichelostemma pulchella—Brodiaea*
   Hemerocallis aurantiaca—Day Lily*
   Zigadenus venonosus—Death-Camas
  FAMILY Agavaceae
   Agave deserti—Desert Agave*
   Yucca baccata—Yucca*

*Additional species discussed in text.

INDEX TO PLANT USES
(Refer to General Index for page numbers)

261

ters, Day Lily, Mariposa Lily, Marigold, Milkweed, Black Mustard, Nasturtium, Milk Thistle, Sow Thistle

Fruits—Beavertail Cactus, Boxthorn, Canaigre, Cheeseweed, Cherry, Chokecherry, Eugenia, Guava, Natal Plum, Fan Palm, Radish (*See also* Berries)

Fungi—Morel, Puffball, Shaggy Mane

Glue—*See* Adhesives

Implements—Acorn, Agave, Barrel Cactus, Bulrush, Cattail, Dogwood, Douglas Fir, Wild Grape, Hazelnut, Ironwood, Juniper, Manzanita, Mesquite, Milkweed, Nettle, Fan Palm, Pine, Rose, Service Berry, Yucca

Jams and Jellies—Beavertail Cactus, Carnation, Chokecherry, Eugenia, Mountain Grape, Wild Grape, Guava, Manzanita, Mint, Fan Palm, Natal Plum, Pyracantha, Raspberry, Rose, Salal, Service Berry, Spearmint (*See also* Berries)

Jewelry—*See* Ornamentation

Juices—*See* Berries and Fruits

Leather Treatment—Canaigre, Ocotillo

Meal—*See* Flour and Meal

Medicinal: Teas—Bay, Blackberry, Canaigre, Celery, Cherry, Chinquapin, Chokecherry, Creosote Bush, Dandelion, Dogwood, Elderberry, Mountain Grape, Horehound, Lemonade Berry, Manzanita, Milkweed, Miner's Lettuce, Mint, Mormon Tea, Black Mustard, Pennyroyal, Peppermint, Rose, Sagebrush, Shepherd's Purse, Squaw Bush, Strawberry, Yerba Santa

Medicinal: Treatments—Acorn, Agave, Alfalfa, Bay, Beavertail Cactus, Blackberry, Bracken Fern, Canaigre, Celery, Cherry, Chia, Chickweed, Chinquapin, Creosote Bush, Dandelion, Dogwood, Elderberry, Wild Grape, Juniper, Lamb's Quarters, Manzanita, Mesquite, Milkweed, Miner's Lettuce, Black Mustard, Nut-Grass, Ocotillo, Onion, Pine, Pinyon, Plantain, Puffball, Purslane, Australian Saltbush, Service Berry, Squaw Bush, Squaw Currant, Squaw Root, Nettle, Sunflower, Violet, Watercress, Yerba Santa

Muffins—*See* Breads and Cakes

Mush—*See* Flour and Meal

Music—Acorn, Elderberry

Nuts—Acorn, Bay, Cherry, Hazelnut, Jojoba, Pine, Pinyon, Walnut

Oils—Bay, Jojoba, Sunflower

Omelettes and Egg Dishes—Alfalfa, Beavertail Cactus, Filaree, Sheep Sorrel, Sow Thistle, Violet, Yucca

Ornamentation—Cherry, Desert Holly, Fir, Mesquite, Pine

Paint—*See* Dye

Pancakes—*See* Breads and Cakes

Pies and Pastry—Acorn, Blackberry, Blueberry, Canaigre, Currant, Elderberry, Gooseberry, Pineapple Guava, Salal, Sheep Sorrel, Walnut (*See also* Nuts, Berries, Cookies, Breads and Cakes)

GENERAL INDEX

Abies concolor, 6
Abyssinians, 148
Acacia, 108-110
Acacia greggii, 108-110, 258
Acer macrophyllum, 6
Acorn, 261-263
  leaching of, 16
Acorn Roca Bars, 69
Adenostoma fasciculatum, 5
Adhesives, 261
Africa, 152, 224, 236, 246
African Marigold, 243
Agave
  deserti, 99-101, 260
  shawii, 100
  utahensis, 99
Agave, Desert, 99-101, 138, 248, 261-
  263
  Sautéed Buds, 101
Agave Skipper Butterfly, 100
Aleppo Pine, 76
Alfalfa, 177, 178, 248, 258, 261-263
  Meal-In-An-Omelette, 178
  Tea, 178
Algaroba, 230-233
Alkali Sink Scrub, 7, 142, 164, 166
Allium
  haematochiton, 69-71, 260
  validum, 70
Alpine Fell-field, 6, 36
Alpine Prickly Currant, 36
Amaranth, Green, 166, 179, 180, 185
  188, 194, 198, 257, 261, 263
  Muffins, 180
  Savory Greens, 180
Amaranthus
  fimbriatus, 179
  retroflexus, 166, 179, 180, 185,
    194, 198, 257
Amatungula, 246
Amelanchier pallida, 83-85, 257
Andes, 179
Animal uses of plants. *See* specific
  animal

Anise, 150, 151
Antelope, 121
Antelope Ground Squirrel, 166
Apache, 99, 127, 132, 144
Apium graveolens, 148, 149, 224, 259
Arabia, 224
Arctostaphylos
  in Chaparral, 5
  glauca, 55, 56, 257
  manzanita, 55
  patula, 55
Aristotle, 184
Arrowhead, 98, 140-142, 146, 260-
  263
  Wappato Salad, 141
  Pancakes, 142
Artemisia
  californica, 7, 134
  tridentata, 7, 134, 135, 259
Ascelpias
  speciosa, 57-59, 259
  syriaca, 58
  tuberosa, 58
Ash-throated Flycatcher, 41
Asia, 43
Asia Minor, 247
Atriplex
  canescens, 115, 211
  confertifolia, 4
  hymenelytra, 115, 256
  lentiformis, 115
  patula ssp. hastata, 166-168, 256
  semibaccata, 211, 256
  tularensis, 115, 211
  vallicola, 115, 211
Audubon Cottontail, 81
Australia, 175
Australian Brush Cherry, 235, 236
Avena fatua, 202-204, 260
Avocet, 143, 171
Aztecs, 104

Backpackers, plants used by, 21, 23,
  25, 33, 45, 60, 81, 85, 96

265